W9-BAE-848

Also by Jay Ingram

The End of Memory: A Natural History of Aging and Alzheimer's
Fatal Flaws: How a Misfolded Protein Baffled Scientists and Changed the Way We Look at the Brain
Theatre of the Mind: Raising the Curtain on Consciousness
Daily Planet: The Ultimate Book of Everyday Science
The Daily Planet Book of Cool Ideas: Global Warming and What People Are Doing About It
The Science of Everyday Life
The Velocity of Honey: And More Science of Everyday Life
The Barmaid's Brain and Other Strange Tales from Science
The Burning House: Unlocking the Mysteries of the Brain
A Kid's Guide to the Brain
Talk Talk Talk: Decoding the Mysteries of Speech
It's All in Your Brain
Real Live Science: Top Scientists Present Amazing Activities Any Kid Can Do
Amazing Investigations: Twins

The Science of Why

Answers to Questions About the World Around Us

Jay Ingram

PUBLISHED BY SIMON & SCHUSTER

NEW YORK   LONDON   TORONTO   SYDNEY   NEW DELHI

Simon & Schuster Canada
A Division of Simon & Schuster, Inc.
166 King Street East, Suite 300
Toronto, Ontario M5A 1J3

This Simon & Schuster Canada edition November 2016

SIMON & SCHUSTER CANADA and colophon are registered trademarks
of Simon & Schuster, Inc.

For information about special discounts for bulk purchases, please contact Simon & Schuster
Special Sales at 1-800-268-3216 or CustomerService@simonandschuster.ca.

Cover and illustrations by Tony Hanyk, tonyhanyk.com
Interior design by David A. Gee
Manufactured in the United States of America

10 9 8 7 6 5 4 3 2 1

Library and Archives Canada Cataloguing in Publication
Ingram, Jay, author
The science of why : answers to questions about the world around us / Jay Ingram.
Issued in print and electronic formats.
ISBN 978-1-5011-4429-5 (hardback).--ISBN 978-1-5011-4430-1 (html)

        1. Science--Popular works. I. Title.
Q162.I55 2016   500   C2016-901996-9
                                    C2016-901997-7

ISBN 978-1-5011-4429-5
ISBN 978-1-5011-4430-1 (ebook)

To Rachel, Amelia and Max

Contents

Part 1: The Body

Part 2: The Animal Kingdom

Part 3: Supernatural

Part 4: The Natural World

The Science of Why

Part 1
The Body

What do our pupils say about us?

PUPILS DILATE (EXPAND) OR CONTRACT AS THE LIGHT DIMS OR BRIGHTENS. But pupils also change size according to what the brain behind them is doing, whether that's recalling memories, analyzing a problem or experiencing strong emotions. We may be unaware that our eyes are giving away so much while our brains are busy, but others who are aware can use that information to gauge their responses to us.

People have been deliberately sending messages with their eyes since at least the Renaissance. Back then, Italian women used eyedrops derived from the deadly nightshade plant—which they called belladonna, or "beautiful woman"—to dilate their pupils, believing that it made them more attractive. It wasn't until hundreds of years later that anyone could figure out why dilated pupils would be so alluring. In the 1960s, a study showed that our pupils dilate when we're looking at something

What is it that the eyes love?

interesting or attractive. So a Renaissance man gazing into the eyes of a woman who had just used belladonna eyedrops would see dilated pupils and unconsciously assume she was looking at something she found appealing: him!

Eckhard Hess of the University of Chicago was responsible for those 1960s experiments, which were among the first to examine pupil dynamics. In Hess's studies, volunteers were shown images on a screen, and a camera photographed their pupils as they dilated or contracted in response to the changing pictures. The light levels were constant from one image to the next, ensuring that the changes in the volunteers' pupils were a response to mental activity rather than to light.

Hess was able to confirm the intuition of those Renaissance women: he found that men judge a woman's face to be more attractive when her pupils are dilated. Even when men were shown the same woman twice—the only difference being the diameter of her pupils—they preferred the image with the bigger pupils. Hess also confirmed that the phenomenon was more general than that. Pupils expanded when an individual saw anything interesting or attractive. But the same person's pupils contracted when he or she saw something unpleasant.

 **? Did You Know . . .** Another study by Eckhard Hess concluded that women who are attracted to "bad boys" (yes, they had a definition for that) responded most positively to males with dilated pupils. And another experiment, this one in the Netherlands, showed that people were more likely to give money to a virtual partner—and thus more likely to trust him or her in general—if that person's pupils were enlarged.

Our understanding of why pupils dilate has improved since Hess's experiments. We now know that pupils dilate in response to a range of mental activities, from recalling memories to making decisions while shopping or playing rock-paper-scissors. And it's not just our pupils that show our brains are at work. Blinking matters, too. Blinks signal the beginning of a mental process. After we blink, our pupils remain dilated as long as we're working on the problem. When we're finished, we blink again as our attention switches to something else and our pupils shrink.

The best data so far suggests that our pupils dilate the most when something is emotionally engaging. It doesn't matter whether that emotion is bad or good, just so long as it grabs our attention. In one experiment, participants filled out a survey asking if they were impulsive shoppers. They then watched a scene of people shopping. The researchers found that the people who identified as impulsive shoppers had the greatest pupil dilation—just viewing the activity of shopping was so emotionally exciting and stimulating for their brains that their pupils expanded.

That tight-knit connection between brain and pupils also happens when thinking is taking center stage. For example, while you try to solve the latest sudoku, you're constantly juggling numbers in your working memory. As your brain is managing those digits, your pupils dilate because of the mental effort. But if you were to stop concentrating and let your mind wander away from the puzzle, your pupils would return to normal.

We sometimes hear anecdotal reports of magicians being able tell what card you've picked out of a deck based on the size of your pupils, or clever shopkeepers knowing what you really want to buy by reading your pupils. But there's little research to support those stories. One experiment that came close to proving these claims used the game rock-paper-scissors to see whether people could predict their opponent's decisions by observing his or her pupils. Instead of playing the game person-to-person, the subjects watched video replays that showed a virtual opponent. Each subject was told that the video opponent's pupils would change when he or she chose an option (rock, paper or scissors). Once the live players had been coached about what to look for, they beat the video players more than 60 percent of the time.

There was one problem with this experiment: because the subjects were watching a replay, they were seeing the video player's pupils dilate after his or her decision had been made. In a live game, players would have to act before that happened, so trying to use an opponent's eyes as a crystal ball wouldn't help most people win an ordinary game of rock-paper-scissors.

Poker, though? That's a different story.

THE END

Why do onions make me cry?

THE BIBLE DESCRIBES HELL AS A GIANT LAKE OF "FIRE AND BRIMSTONE." Brimstone is really just an ancient term for the element sulfur. A lake of brimstone would have a pretty awesomely powerful smell, but you don't have to wait to experience it—look no further than the small but potent onion.

As you cut an onion, the knife blade breaks down the tissues in it, releasing chemicals that normally never come in contact with each other. When they do come into contact, they begin to rearrange themselves into new combinations of sulfur-containing compounds. Ultimately, a molecule named syn-propanethial-S-oxide is released into the air. When receptors in the cornea of your eye sense its presence, they release protective tears to wash it away.

Life stinks.

That tear reaction peaks roughly half a minute after you first cut into the onion, and it takes about five minutes to go away. It's impossible to turn off this chemical reaction, but there are some ways to make the process a little less painful. Some people suggest putting the onion in the fridge or freezer before you cut it. Chemical reactions are slower at lower temperatures, so if the onion is cold when you first cut into it, you might be able to finish the job before the tear-inducing chemicals are released. There have been all sorts of other suggestions, such as cutting the onion while it's under water, covering the lower part of your face with a paper towel, running a fan, dicing the onion under a running stove vent or even wearing safety goggles. So far, nothing has proved absolutely foolproof—the onion always wins.

Do we all have Neanderthal in us?

I LOVE THE NEANDERTHALS, and apparently if I had lived several millennia ago, I might have meant that literally! Neanderthals are the other human species, the "cavemen" who lived in Europe from about 350,000 to 40,000 years ago. Neanderthals have typically been portrayed as primitive, clumsy, heavily built, knuckle-dragging, unintelligent brutes who had scraped out an existence in Ice Age Europe before dying out when confronted by the graceful, more intelligent, craftier modern humans.

It's true that Neanderthals were, on average, stockier than modern humans. They were muscular and powerful, built for rapid movement from side to side. Compared to modern humans, Neanderthals looked quite different. They had relatively large noses—in fact, the nose and cheekbones were slightly pulled forward compared with ours. Their larger sinuses

Evolution Makeover

might have been designed to warm inhaled air, a necessary feature in a cold-climate hominid. Their brains were bigger and heavier, and they had a different shape, with a prominent bump at the back. It's been speculated that this was an enlarged cerebellum, the part of the brain that coordinates movement, among other things. That would make sense, given that much of the Neanderthal brain space was devoted to vision and muscular movement, with less allotted to the frontal lobes, where our ability to plan and make decisions resides.

In a perfect example of how a single discovery can set an inaccurate tone for nearly a century, a Neanderthal skeleton excavated in France in 1908 was described as being hunched over, more animal than human. Years later, closer study of the remains revealed that the excavated individual had been crippled by arthritis in his spine. It was the arthritis, then, and not some characteristic feature of his species, that accounted for his inability to stand erect.

The new information didn't help, as the Neanderthal's unimpressive résumé had been set in stone. The old saying that the victors write the history books has never been truer than for the Neanderthals. At first, the fact that we were still here and they weren't seemed to tell the whole story. Shortly after they encountered modern humans (us), the Neanderthals slowly withdrew to a few final tiny outposts in southern Spain and Portugal before disappearing for good, and that the Neanderthals didn't last was thought to prove their inferiority.

Did You Know . . . Neanderthals have a sister group, called the Denisovans, a mysterious, newly discovered group that lived around the same time as the Neanderthals—tens to hundreds of thousands of years ago—and apparently spread their genes around, too.

In the last twenty-five to thirty years, though, the Neanderthals have undergone a radical image rehabilitation, to the point that what made them different from us seems much less obvious than what made them the same.

Culturally, Neanderthals have always seemed a bit backward compared to modern people. Our ancestors—the modern people of the Neanderthals' era—are responsible for most of the thirty-thousand-year-old cave paintings in southern France and Spain, but they don't have a monopoly on art. There is an example in Gibraltar called Gorham's Cave, where researchers have found tic-tac-toe-like scratchings on a wall—abstract art, some call it—that date to when the cave was occupied by Neanderthals. (Gorham's Cave is apparently one of the last places they lived.) And it's not just the cave art that hints at culture: there's solid evidence that Neanderthals buried their dead, used fire and decorated their bodies with teeth, claws, feathers and red, black and yellow pigments. They also hafted blades onto spears and were capable of crossing open water (presumably voluntarily).

The Neanderthals were hunter-gatherers, and so they had to go where the food was. Evidence of the Neanderthal diet is gleaned from chemical analysis of micro-traces of food left behind in the plaque on fossilized teeth. Neanderthals consumed a wide-ranging diet that was dominated by meat—there are butchered animal bones aplenty wherever their remains have been found—but that also included vegetables. Chemical analysis of a famous Neanderthal skeleton from Saint-Césaire, France, suggested that he and his cohort ate fewer reindeer and more woolly rhinos or mammoths than did their main competitors, hyenas. The rest of their potential prey—deer and horses—they split fifty-fifty with the hyenas. There are also hints of Neanderthals consuming some plants for medicinal purposes.

The Neanderthal Diet

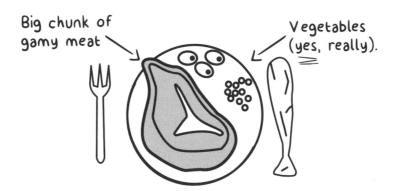

Big chunk of gamy meat

Vegetables (yes, really).

Canadian scientist Valerius Geist came up with the most insightful and spectacularly visual account of how Neanderthals managed to hunt and kill the biggest mammals of their time: rhinos and mammoths. The Neanderthals' spear points were coarse and heavy, unlike the finely shaped versions preferred by our ancestors. The Neanderthals' craftsmanship has often been seen as a deficiency attributed to their supposedly primitive skills, but Geist argues that those clumsy spearheads were perfect for the job. He claims that Neanderthals worked at close quarters when hunting, and that only two men were needed. One Neanderthal would approach the animal from the front (carefully) and irritate it. At the same time, hunter no. 2 would close in from the side and grab the mammoth by the fur. (Mammoths and rhinos of the time had long hairy coats, and Neanderthals had exceptionally strong hands with long fingers and strong, broad fingertips.) The animal, disturbed by the thing clinging to its side, would wheel and buck, trying to shed the nuisance. Hunter no. 1 would take advantage of the distraction to drive his spear into the animal. The lighter, slimmer spears preferred by our ancestors were good only for throwing; in close quarters, they likely would have broken off on a bone. Two Neanderthals working right next to the animal, then, could do the work of five modern humans throwing from a distance.

Geist compares the Neanderthal hunting style to the way that rodeo clowns distract bulls. In fact, the pattern of bone breaks seen in Neanderthal skeletons mirrors closely the typical injuries suffered by rodeo clowns today. A study in 1995 compared bone breaks among modern humans living in New York and London, Native Americans from hundreds to thousands of years ago, Neanderthals and rodeo clowns. The Neanderthals and the clowns stood out from the rest in that they both had a much higher percentage of head and neck fractures and fewer lower-limb breaks. It's not proof, but it's pretty suggestive that Geist—who got his idea of close-encounter hunting while watching bull riders and clowns at the Calgary Stampede in 1973—might be right.

Whether or not Neanderthals used language is a long-running controversy. Earlier theories argued that they died out because they didn't have the brain machinery for language, spoken or signed, and therefore couldn't compete with the more highly developed modern humans. One key piece of that argument was that Neanderthals' throats didn't appear to have a bone, the hyoid, that plays a crucial role in articulating speech. But—in yet

another example of how the picture suggested by advancing science is always changing—a sixty-thousand-year-old Neanderthal skeleton found in Israel apparently has a hyoid identical to ours. That doesn't mean that the individual, or its species, could speak, but at least there's one less barrier to that possibility. Unfortunately, speech leaves no fossil traces.

To understand whether modern humans have any Neanderthal in them, we must turn to genetics. The technology that allows us to reconstruct large portions of the Neanderthal genome is staggering and brilliant. The first sets of Neanderthal genes were extracted from a fraction of a gram of bone! Scientists have even been able to use genetic analysis to show that some Neanderthals would have been fair-skinned redheads.

The bottom line is that, yes, in those of European descent, up to 5 percent of the genome can be Neanderthal genes. Five percent of the modern European genome sounds like a lot, but remember that a majority of those Neanderthal genes probably code for things we're never aware of, like some tier-2 protein in the kidney. It's tempting to think of the Neanderthal contribution as the "gene for the big nose" or "the gene for muscular legs," but so far there is no evidence that we have Neanderthal genes that tweak physical differences.

That's not to say that Neanderthal genes aren't useful. There's a cluster of them on our chromosome 3, and some of those play a role in adapting to the ultraviolet light in sunlight. Those particular genes are even more common in East Asians' genomes. Somehow, modern people acquired a set of UV-light-adapted genes outside of sunny Africa that were more effective than the ones they'd had upon leaving it. A second, vital set of imported Neanderthal genes is found in the HLA system, a group of genes that gives the immune system the ability to identify and defend against invading bacteria and viruses. The protein molecules produced by HLA proteins can confer resistance against infections, including the Epstein-Barr virus, which causes infectious mononucleosis and is associated with a cancer called Burkitt's lymphoma.

Neanderthal genes are even more common in East Asians, but they're effectively absent in Africans. This has led to the conjecture that the interspecies hanky-panky happened mostly in the Middle East, especially the Levant, the area around the east coast of the Mediterranean. We'd all like to know what those gene-swapping encounters were like. It's well known that the human brain is very good at evaluating others as "us" or "them," and that reaction had to be heightened when modern humans met the Neanderthals. But did the reaction intensify the exotic or tune it down? Was sex consensual? So far, the science doesn't allow us to say definitively whether the matings were between Neanderthal males and *Homo sapiens* (us) females or vice versa, but the former seems most likely. That's partly because in this case, males would be more likely to be infertile, leaving no trace of themselves, and also because genes that would be unique to female Neanderthals—mitochondrial genes—are absent in modern populations.

And how often might these interactions have happened? We don't know. I've seen numbers ranging from one mating per several dozen encounters—a mere handful per year—to totals of several hundred or even a few thousand. Not an orgy, but apparently often enough to leave behind evidence that has lasted more than forty thousand years.

Whatever the exact number is, one thing is certain: the notion of a Neanderthal dragging any woman back to his cave by the hair is nothing but myth. What sort of evidence would one need to be able to believe this? A stretched-out skeletal hand with long hairs in its grasp? Consistent sets of micro-fractures on female Neanderthal skulls? There isn't any such evidence.

The earliest reference to this idea is from the nineteenth century. One Andrew Lang, a prolific and wide-ranging writer and commentator, wrote about nomadic life in Europe just as the ice sheets were retreating. He had this to say about that period: "In the big cave lived several little families, all named by the names of their mothers. These ladies had been knocked on the head and dragged home, according to the marriage customs of the period, from places as distant as the modern Marseilles and Genoa."

Out of the mind of an imaginative man directly into the public awareness. Lang didn't specify that this description was about Neanderthals—though they had been discovered by then—but once the Neanderthals became popular, it was easy to fit them into that scenario. It took many years to dispel the notion and prove that it's just the stuff of cartoons.

?

Did You Know . . . Neanderthals have traveled across time and space in pop culture. English author William Golding is best known for his first novel, *Lord of the Flies*. His second book, *The Inheritors*, described an encounter with and subsequent elimination of a small group of Neanderthals by a band of modern people. His Neanderthals were a curious breed, guided by pictures in their heads, incapable of much language and perplexed by the bows and arrows used against them. Of course, Golding was speculating (this was 1955), but although he was likely wrong about the archery, there is evidence that Neanderthals had larger eye sockets and perhaps were a more visual and less linguistic people than we are.

Years later, science fiction author Robert J. Sawyer wrote the Neanderthal Parallax, a trilogy in which the Neanderthals survive in a parallel universe. Eventually, one individual somehow crosses between these universes to meet modern humans at the Sudbury Neutrino Observatory, in Ontario. He would be the only Neanderthal ever to know anything about quantum physics—as far as we know.

Even after 160 years of study, no one is really sure why the Neanderthals died out after managing to exist across Europe and the Middle East for hundreds of thousands of years. The traditional view—now largely abandoned—was that modern humans killed them off (admittedly a very human thing to do), and it's true that the timing of our ancestors' arrival in Europe is suspicious: the latest dating techniques suggest that modern people arrived in Europe about forty-five thousand years ago, and the Neanderthals, although already well established, were pretty much finished five thousand years later. True, five millennia is a long time over which to extinguish a species, but both populations were sparsely distributed, so contact might have been rare. The problem with this theory is that there's no fossil evidence of pitched battles between the two closely related hominids, and although modern people supposedly invented both the throwing spear and the bow and arrow, you'd think that hand-to-hand battles would favor the more physical Neanderthals. Some have conjectured that it was tropical diseases we brought from Africa that did the deed.

But the preferred explanation these days points to a combination of environmental changes. As the glaciers continued to recede and their favorite prey dwindled, the Neanderthals found themselves unable to cope with those changes and also compete with us moderns. But although the Neanderthals as a species disappeared, the fact that a portion of their genome clings to ours shows that their rodeo-clown legacy is still alive and well.

Why does asparagus make my pee smell funny?

IF YOU'RE ONE OF THOSE PEOPLE who can identify a distinct aroma in your urine after you've eaten asparagus, then you're one of the lucky ones! Well, if not lucky, then you're at least among those who are able to smell it.

Over the years, many of the chemicals found in asparagus have been accused of causing the unusual odor, which can affect a person's urine when just a few spears are eaten. Although scientists had various culprits in mind, they did agree on one thing: something containing sulfur caused the asparagus smell.

But each time scientists thought they'd identified what those chemicals were, they were unable to explain how they could survive the long trip down the human digestive tract. The journey from fork to bladder was more than enough time for those compounds to be broken down and absorbed by the body, so post-asparagus urine should be

I'm stalking you.

scent-free. If those chemicals are still present in a person's urine after dinner—and we know they can't have survived their trip intact—then it must be that some larger molecule survived and at the last minute changed into the odor-causing sulfur-based substance.

The best candidate for that large molecule is the appropriately named asparagusic acid. While asparagus is still in the ground, this acid serves as the plant's defense against nematode worms. There's more acid in the asparagus when it's young and still growing, which might explain why the pee odor is so much more pungent after you eat a young tip than an older, tougher stem. But the full details of the chemistry of the asparagus odor have yet to be worked out.

That's the accepted explanation for how asparagus gives urine that odd smell. But here's the catch: not everyone creates the scent. Experiments since the 1970s have produced conflicting results about how much of the population can produce this asparagus smell, with estimates ranging anywhere from 40 to 90 percent.

Similarly, not everyone has the ability to smell the stench. In one study, people were exposed to two samples of urine, each of which contained traces of asparagusic acid. The first set was original pee samples, while in the others, the asparagus scent had been diluted to an incredibly small 1 in 4,096 parts. A number of people in the study were unable to detect any stink in the undiluted urine, suggesting that they had no ability to smell the sulfur, while 10 percent were able to detect the asparagus smell even in the diluted sample, showing that they were hypersensitive to the odor.

A 2010 experiment added some detail. Volunteers ate a few stalks of asparagus and after a few hours submitted a urine sample. A couple of days later, the participants ate a piece of bread as a control substance and again deposited their urine. The researchers then recruited urine-sniffing volunteers, had them do their thing and produced the most precise numbers yet: in roughly 8 percent of the asparagus samples, there was no sulfur smell at all. And when the smell was there, 6 percent of the volunteers couldn't detect it.

It's just that asparagus makes my urine smell funny.

Asparagus turned my humble chamber pot into a bower of aromatic perfume.

The biggest discovery from the experiment wasn't how many people were able to produce or smell the asparagus odor; it was how they were able to do so. The researchers found evidence of a gene that determines whether a person has the ability to smell asparagus in urine. As we've just seen, earlier experiments had shown that there's a wide range of sensitivity to the smell—to some people, the stench is overpowering, whereas others don't notice it at all. Knowing that, the scientists in the most recent experiment concluded that there couldn't be a single asparagus smell. Rather, there must be a variety of odors, each one as different as the person who produces it. That has yet to be tested, though—it's hard enough to persuade volunteers to sniff urine at all, let alone to judge its quality.

Why do I sometimes have difficulty recalling words on the tip of my tongue?

YOU ARE A RARE PERSON IF YOU HAVEN'T EXPERIENCED THE tip-of-the-tongue phenomenon—that's when you're trying to remember a word or name and it feels right there, "at the tip of your tongue," but you can't quite access the information and spit it out. Words usually spill out of our minds and mouths at incredible speed without much mental effort, so it can feel awkward and frustrating when we suddenly can't recall the simplest thing. It's estimated that when you're in your twenties, you will experience this phenomenon about twice a week, but forty years later that rate will have doubled.

Damn it, I
can't remember
her name.

It'sss on the
tip of my tongue.

Hey, boysss...

If you've experienced this phenomenon, you can attest that it provides a small, short-lived window into your own brain. It's like waving a flashlight around inside your head, trying to illuminate a word that you know is waiting there but that is playing a game of hide-and-seek. And not only do you know that pesky word is in there, you even have a sense of its size and shape. But you still can't find it.

In 1893, the great American psychologist William James noted that when we're searching for a word and the wrong ones are offered to us, we know that they're not what we're looking for, that they don't fit the mold. But that doesn't solve the problem.

So why are psychologists like James interested in this experience? Because it's like seeing a hummingbird's wings in slow motion: an opportunity—maybe—to study the processes of memory and word-making more closely. The next time it happens to you, take a moment to savor the feeling and explore it instead of pulling out your hair and fearing for your sanity.

 TRY THIS: Want to experience tip-of-the-tongue effect right now? Try naming all seven dwarfs and watch what happens.

Here's what you might find: the difference between a tip-of-the-tongue experience and straightforward forgetting is that with the former, you know you know the word you're looking for. It fits the sentence and expresses what you want to say. In that sense, it's much closer to the surface than a fact that you have completely forgotten and become aware of only when you're reminded.

Research shows that you likely know a lot about that lost word even if you can't dredge it up. The classic tip-of-the-tongue study that revealed this (and set the stage for the modern approach to the subject) was published back in 1966, by Harvard psychologists Roger Brown and David McNeill. They gave students the definitions of forty-nine uncommon words, defined as occurring less than once in a million words in common usage. (They didn't include definitions of "super-rare" words, ones you come across every four million words.) They did not supply the words, just the definitions. (Whether these words are as common today, fifty years later, is arguable, given that they included ambergris, caduceus, nepotism, fawning, unctuous, cloaca, sampan and philatelist. Don't worry: if you don't know the meaning of most of these, you are not alone!)

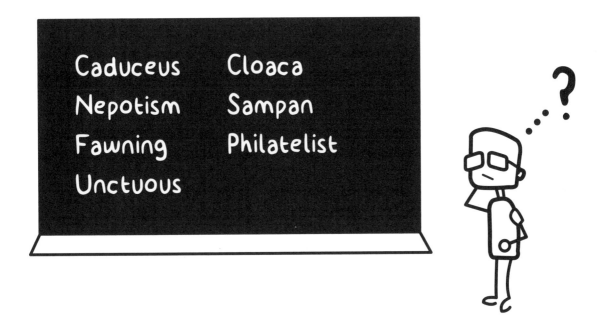

The psychologists then asked the students to come up with the right words to match the definitions. And the students came through, experiencing countless tip-of-the-tongue moments, memorably described by Brown and McNeill as appearing to be in "mild torment, something like the brink of a sneeze." When students thought they knew the word but couldn't immediately verbalize it, Brown and McNeill asked them to list any features they thought they knew, such as the number of syllables, the first letter, and words of similar sound and similar meaning. The intriguing thing was that the students were able to do quite a bit of that.

Guesses at the number of syllables were impressively accurate. For instance, when people were searching for three-syllable words, two-thirds of them guessed the mystery word was indeed three syllables. The same accuracy rate held true for shorter words, but it was lost for words longer than three syllables.

Sometimes the students were seeking the wrong word, but when they were right, they were able to guess the first letter accurately 57 percent of the time. For example, when the students were told the word was defined as a flat-bottomed boat common in Asia and propelled by oars (a sampan), those experiencing tip-of-the-tongue effect recalled similar-sounding words like Saipan, Siam, Cheyenne, sarong, sanching and sympoon. (The last two are not actual words, by the way.) The students were often able to come up with words with similar meanings, such as barge, houseboat and junk—all of which make sense but are more common words and are not physically or structurally like sampan.

Perhaps the most surprising result of this experiment was that when words of six letters or more were the target, students did well at guessing letters at the beginning and end of the word, but poorly when it came to letters in the middle. This led Brown and McNeill to the image of the mystery word as the tall man in the bathtub: his head sticks out at one end, his feet the other, but his middle can't be seen.

Explaining these results in terms of how words are stored in our brains is tricky. Brown and McNeill argued that, yes, words are stored in the brain in their entirety but can be retrieved without having to know every single letter. That was made clear by the students' correctly guessing the first and last letters in the experiment, but it's also true when

y— — r— —d th— l—st f—w w—rds —f th—s s—nt—nc—.

E_rek_!

The reason you feel the tip-of-the-tongue phenomenon so strongly is that you are aware of the meaning and appropriateness of the word but just can't connect those things to the actual letters. It might happen because you don't use the word often (or haven't used it for a while) and so the connection has gradually weakened. And age is a factor, too.

Since this classic 1966 study, there have been dozens of attempts to clarify what sorts of cues might help resolve a tip-of-the-tongue quandary. One of the most interesting was conducted by Lisa Abrams and Emily Rodriguez in 2005. They were interested in what sorts of words would help relieve the tip-of-the-tongue state. So with the question "What do you call a large colored handkerchief usually worn around the neck or head?" they were looking for the word *bandana*. If one of their subjects ran into a tip-of-the-tongue issue, he or she was asked to read from one of three word lists. One list included a word with the same first syllable, such as *banjo*; the other list included a word that had the same first syllable but was a different part of speech, such as *banish*; the third list had unrelated words.

What Abrams and Rodriguez found was that *banish* helped but *banjo* didn't. Why? They concluded that words that are the same part of speech as the target word (in this case, the two nouns *bandana* and *banjo*) compete in the brain, whereas *banish*, a different part of speech stored in a different part of the brain, helped recall.

Did You Know . . . Synesthetes—people whose sensory experiences cross wires—can hear colors or taste words. When synesthetes experience tip-of-the-tongue effect, they report tasting the word they're looking for before they know what it is they are tasting!

One very cool new addition to this study is an online diary where you can record your tip-of-the-tongue experiences, which are then made available to researchers. Spedi—or Speech Error Diary—is run out of the University of Kansas. Citizen science at its best!

Do fingernails grow faster than toenails? If so, why?

Your fingernails, including the thumbnail, all grow at roughly the same rate, usually about 0.1 millimeters a day, or just under a millimeter a week. Toenails grow only half as fast, averaging a mere 0.05 millimeters a day, or less than half a millimeter a week. There's the first question answered—that was easy! But to understand exactly why fingernails grow faster than toenails, we need to examine how our nails grow in the first place.

Take a look at your thumbnail. At the farthest end, the nail loses contact with the finger and launches itself into space, providing an opportunity for dirt and infectious organisms to lodge themselves underneath. But as you move down the nail, you'll notice that the skin of your thumb surrounds and grows over it on both sides and at its base, forming a protective layer that prevents anything from getting into the soft tissue below.

Cast your eyes down a little farther, where the base of the nail meets your skin, and you'll see the white semicircle known as the lunula (the word means "little moon"). The lunula and other tissues around it, like the fleshier-colored matrix, create the nail by continuously churning out cells; the fastest-growing skin cells anywhere in the body, in fact. Most of the nail is generated by the cells closets to the luluna. Those cells change, harden and eventually die as they move towards the tip of the nail, leaving behind nothing but keratin the main ingredient of the hard-shelled nail itself.

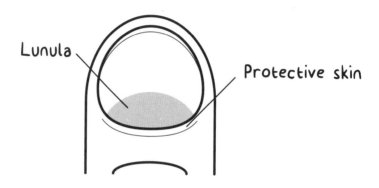

Lunula

Protective skin

? Did You Know . . . Have you ever noticed that your nails tear more easily across than down? Three layers of material make up the nail. The top and bottom layers are weak sheets of keratin that can tear in almost any direction. (They might not be strong, but they likely enhance the nail's ability to bend without breaking.) The middle layer is the strongest—six times tougher than the other two layers—and the keratin in it lies crossways, so it's four times easier to tear across than from the tip to the nail bed. And that's why your nails are easier to tear from side to side.

We all know that if you bash your thumb, the damage done will soon reveal itself as a mark on the nail, and that blemish gradually migrates out toward the tip of your thumb as the nail grows. Sometimes, if there's a tiny amount of bleeding, the mark left is what's called a splinter hemorrhage, a tiny straight line pointing toward the end of the thumb. These formations actually tell you something about the growth of your nail: the underside of the nail is striped with grooves that match similar channels on the surface of the tissue beneath the nail.

Given how nail growth works, it makes sense that damage to a nail can compromise its growth. In the early 1980s, a dermatologist in England, Rodney Dawber, suffered tendon damage to his left ring finger during a rugby match and treated the injury by putting his damaged finger in a splint. In true nerd style, rather than bemoaning his bad luck, Dawber experimented with his injury. He was curious to check out the idea of "terminal trauma," the negative impact of a finger injury on nail growth (even if that injury didn't damage the nail itself). For the three months he had the splint on his finger, he compared the growth rate of the nails on his two ring fingers, and he found that the growth on the injured finger slowed by about 25 percent. He also discovered (although this had been noted by others) that the nails on his dominant hand (his right) grew faster than those on his secondary hand. Meanwhile, his right and left toenails grew at about the same rate.

Dawber's experiment raises interesting possibilities, but it still doesn't definitively answer why some nails grow faster than others. There are just too many reasons why the growth of his injured finger's nail might have slowed. The so-called terminal trauma effect could have been the result of a reduction in circulating blood in that finger, or even simply the lack of stimulation that accompanies activity. To get a better idea of what factors affect fingernail growth, we need to travel back to before Dawber jammed his finger in a rugby scrum. Because, really, everything you need to know about the growth speed of fingernails can be traced back to one man: William Bennett Bean.

Bean kept a continuous record of the growth of his own nails from 1942, when he was thirty-two, to 1977, at age sixty-seven—thirty-five years! His work wasn't in vain, either. Through Bean, we learned that fingernails grow fastest when you're young and slow down over time. Warmth seems to stimulate the growth of nails, so if you live in a hotter climate, your nails will grow faster (although some recent studies have contradicted this). You'd think that would mean Canadians as a whole should see their nail growth speed up in summer and slow down in winter, but as Bean discovered, if you're working in a climate-controlled office that's warmer in winter and cooler in summer, the growth rate won't vary much. He also relayed information he garnered from other sources. For instance, he learned (à la Dawber) that immobilizing the arm or hand—such as by wearing a cast—will slow the rate of fingernail growth significantly, whereas being pregnant speeds growth. Bean even provided another angle on the Dawber injury theory. His own nail growth slowed considerably while he was ill with the mumps, showing that even indirect physical trauma can affect nail growth.

Bean presented all this data objectively—as is proper for a scientist, of course—but he admitted that watching his nail growth slow down over time hit him hard emotionally. (Dawber expressed exactly the same distress.) For Bean, it was the most obvious sign that he was getting old. While he could convince himself that he looked pretty good in the mirror, his nails were telling him something else.

 **Did You Know . . .** Climate affects not only nail growth but also hair growth. In 1917, the dermatologist Felix Pinkus started recording the growth of a single hair on a mole on the back of his hand. The mole's hair would grow for a time before falling out. The hair regrew fourteen times over nine years. The hairs differed in lifetimes from 107 to 195 days—a huge disparity—and summer hairs lasted longer than winter hairs by an average of a couple of weeks. (Eventually, the mole changed and the hair disappeared for good.)

Now we come to the crux of the matter: Why do fingernails grow faster than toenails? It seems that they must continually be refreshed because they incur a lot of constant, if subtle, damage. The list of activities that might damage your nails is, at first glance, pretty slim. It's rare that we sink our nails into something or find ourselves clinging to a tree trunk by them. They're not talons or claws. But if you doubt that you're using your nails all the time, set an alarm and note what you're doing when it goes off. There's a good chance it involves using your nails. Even if you're typing, the nails reinforce the tip of the finger when it pushes down on a key.

There are even verbs in English that specifically describe actions you do with your fingernails: think of *scratch* and *pick*. If you're looking for more evidence, look no further than the fact that the nails on your dominant hand grow a little faster than those on your secondary hand. This tells us that healthy, active, well-nourished tissue is the basis for fast nail growth. Or consider that your middle finger is the longest and therefore the most exposed, and indeed its nail grows faster than the others.

By comparison, the toenails are pampered. For much of their active life they're in a protective shoe or slipper. Tapping or wiggling are the best bets for sustained, regular activity. Toenails rarely have to hold a pen or tie shoelaces, so if faster growth rate is a result of more frequent activity, there's no contest.

Another possible influence on the growth of nails is exposure to the sun. If that's true, toenails will indeed grow slower, given that they rarely glimpse the sun. You could argue that feet see less sun simply because of where they are on the body and the fact that they're so often shaded. Yet even in sunny parts of the world where toenails live a much brighter life, there isn't much evidence that they grow at the same rate as fingernails. And what data exists doesn't point to any difference between feet that are sun-exposed and those that aren't.

A twist on this idea is that a nail's location on the body counts. Blood circulates more easily out to the arms than it does down to the feet, as returning blood to the heart from the legs is a much more difficult fight against gravity. The fact that it's more taxing for blood to circulate through the feet might also explain why toenails grow slower than fingernails. It's pretty much agreed that circulation is a factor, but the test of this theory would be to see whether such a diminished supply of fresh blood to the toes could be responsible for a nail growth rate that is less than half that in the fingers.

There is a tricky part to all this, too: toenails grow slower than fingernails, but they also grow thicker. The most common explanation for this is that toenails are constantly being punished by socks, shoes, kicking doors open and the like, putting the same sort of general stress on our toes that our fingers experience. But one question inevitably leads to another: If the constant activity and micro-injuries that are happening to your toes are similar to those happening to your fingers, why don't toenails grow quickly like their digital colleagues? Why thicker instead?

Science Fiction! It's a myth that our nails continue to grow after we die. A scientific journal (although admittedly not a top-of-the-line one) once ran a study in which the author produced data showing that nails in cadavers continued growing for anywhere from eight to ten days! This has long been a popular horror-movie scenario, but no credible expert believes the results of that study. The apparent growth of nails after death is actually caused by the drying and retraction of the tissue around the nail, making them look longer than before. That puts a nail in that coffin, once and for all.

Why do some people faint when they see blood?

A LOT OF SCIENTISTS WILL ANSWER THIS QUESTION BY SIMPLY SAYING, "Nobody has a clue." That might be short and to the point, but it's far from satisfying. There are answers, but they're contentious, and one in particular has the uncanny ability to make some scientists' blood boil.

Lowered blood pressure, an irregular heartbeat or low blood sugar can all result in a momentary loss of consciousness called fainting—or in medical terms, syncope (rhymes with "canopy"). Humans can faint for all sorts of odd reasons—you may hear of people fainting after coughing, after urinating or after stretching. Sometimes, you can faint simply by getting out of a chair too quickly. In most cases, standing up suddenly causes blood to pool in the legs, lessening flow to the brain. If that dip in blood pressure is extreme, it knocks you out for a short time. You recover because blood flow is re-established to your brain, either because you fall down or because you have the presence of mind to tuck your head between your legs just before you black out.

In both cases, your head is positioned below, or at least no higher than, your heart, so refilling the brain with blood is easier.

But all these versions of fainting are straightforward physiological events triggered by physical stimuli, not mental conditions. How do we explain what happens when fainting is induced by pain, anxiety, emotional stress or fear? Some people faint at the sight of a needle. We assume that's brought about by the expectation that pain will soon be inflicted, but that fear has little connection to the physiological consequences of, say, standing up too quickly. Instead of explaining these cases with physiology, we turn to evolution.

In 2005, Rolf Diehl, at the Krupp Hospital in Essen, Germany, suggested that fainting, with its accompanying drop in both blood pressure and heart rate, was a protective response exhibited by wounded animals. He reasoned that if an animal starts bleeding (and we are animals, remember!), its initial response is to constrict blood vessels and jack up blood pressure and heart rate, so as to maintain circulation in the face of blood loss. But if the bleeding doesn't stop and the animal's blood loss reaches a critical value—roughly a third of its total blood volume—the reverse kicks in: blood vessels loosen, the heart rate drops and the animal's circulatory system slows until it eventually loses consciousness. (This isn't the same as playing dead. An animal playing dead still has its heart pumping and its nervous systems on high alert, exactly the opposite of being in a faint.) But while passing out may leave an animal vulnerable, Diehl argued that the drop in both blood pressure and heart rate buys precious time, allowing blood to clot and thereby reducing total blood loss. If the animal maintained a normal blood pressure in this critical situation, he argued, it would actually hasten death, not prevent it.

What's fascinating is that this straightforward physiological mechanism in humans happens not only in response to the loss of one's own blood but also to someone else losing blood. You'd think the emotion in play here would be fear—seeing someone else's blood raises fears that you'll see your own depleted next, and so your body shuts down and prompts clotting. But Diehl found that in cases where people fainted at the sight of blood, the feeling was one of disgust rather than fear. That feeling was followed by lowered blood pressure, slower heart rate and passing out. If the body's response to heavy bleeding makes survival sense, the mind's response to disgust in these situations certainly does not—falling into a dead faint at the feet of a predator because you're disgusted is not a sensible survival strategy.

Did You Know . . . Human beings have a long history of reacting with disgust to bad smells, such as feces and rotting flesh. Not only that, our disgust reflex has evolved to be a reaction to unpopular politicians, unpunished criminals and even a jacket once worn by Hitler.

The most interesting twist on this already peculiar reaction is that in the doctor's office, the disgust and fear that people feel is directed at the needle itself, not at the blood that it might draw. A needle might suggest the looming presence of blood, but really, most injections are close to 100 percent blood-free. So, something more must be at work.

One psychiatrist, Stefan Bracha, has proposed a controversial theory about needle fear. Bracha claims that this particular phobia dates back to Pleistocene times—as much as two hundred thousand years ago—an era he characterizes as extremely violent. Archeologists have found definitive evidence that tribe-on-tribe and band-on-band disputes were frequent and gruesome. The weapons of choice were typically spears and axes. Deaths were gory and crude. The mere sight of blood was bad news. At some point in those conflicts, the best opportunity for survival—at least for

I get the point. Nobody likes me.

noncombatants, such as children and young women—might have been to faint. A body lying in a heap on the ground could easily be overlooked, while one still upright remained a prime target.

A dig that began in 2012 on the shores of Lake Turkana in Kenya found evidence of a ten-thousand-year-old massacre, where it was obvious that sharp, pointed weapons like arrows and spears had been used in the slaughter. Of twelve almost-complete skeletons, ten had clearly died a violent death. Of course, this isn't proof of Bracha's thesis, but it does at least set the scene. If this fainting response has been passed down through generations, it might be genetic. At the same time, this hypothetical gene can't have been too widespread. Having entire tribes collapse on the ground at the sight of a spear would have been suicidal, if not genocidal. So if the gene exists at all, it is probably present in only a minority of people.

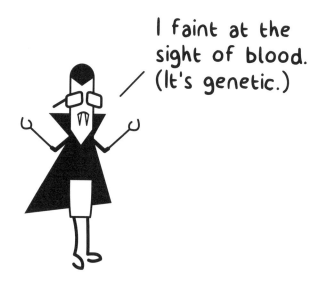

I faint at the sight of blood. (It's genetic.)

Bracha's argument assumes that this reaction is a human-only phenomenon, so the first time that a gorilla or chimp faints at the sight of a needle, the theory will have to be tossed out. So far, not surprisingly, no such experiment has been conducted. It's also true that there is more fainting when an expert (say, someone who's been doing it for thirty years) draws blood rather than someone less experienced. Apparently, the expert wastes no time coddling the patient and therefore appears to be more threatening than the less experienced practitioner.

Theories like Bracha's make some scientists feel faintly uncomfortable. The broad field into which Bracha's theory falls is called evolutionary psychology, and its proponents have been accused of applying genetic explanations to modern-day phenomena when there's no justification for doing so. Take, for instance, the doubtful claim I once heard that the reason girls dream of monsters under the bed and boys dream of monsters coming in the window is that back in the days of our australopithecine ancestors, females roosted in trees at night while the males slept at the base; danger, therefore, came from different directions. There's zero evidence for any part of this claim, though it is entertaining. While Stefan Bracha's theories remain just that, he has at least applied evidence to a puzzle that otherwise lacks explanation.

Why do people choke under pressure?

Most of us associate choking—underperforming when success is expected—with sport. An athlete has a big game or competition, and despite years of training, when the moment comes for the star to shine, he or she fails—sometimes extraordinarily. Choking can happen in the classroom or the boardroom, too, but the sports examples seem the most vivid: baseball player Bill Buckner fumbling a routine grounder in 1986 or pro golfer Greg Norman at the Masters in 1996, leading by six strokes on the final round, then losing by seven in the end.

Psychologists tell us that as incentives for winning increase, people perform better—but only up to a point. After that point, performance starts to decline even as the incentives continue to rise. Two popular but opposed explanations are that choking is the result of either a distracted mind or the opposite: a mind paying too much attention. A third possibility is a mind so excited, so on edge, that the nervous system cannot function at the level it must.

Why does this happen?

Let's use putting in golf as an example. Whether you putt at the golf club or at the mini-putt, you have probably experienced that moment when the pressure is on. Maybe a player has issued a challenge; maybe there's beer on the line. You have an easy shot—one you've done many times before. You prepare, you putt . . . and the ball stops three feet short of the hole. You haven't just missed—you've choked. You could have made that putt in your sleep!

Actually, that is the point of one of the theories: with actions that you have practiced and made automatic, interventions from your conscious mind hinder rather than help performance. Too much "top-down" brain activity is, at these very crucial times, the last thing you want. The unconscious mind has done the training and will serve you better. Too much focus can dismantle a familiar routine into too many pieces—like aiming, shifting, aiming again, rethinking when putting. One hint this might be a true picture of the situation is that experienced putters do better when they have less time for each putt (they've already developed an unconscious putting routine), whereas novices benefit from any extra time.

The alternative theory is that distraction, rather than overthinking, causes choking. Imagine extraneous information pouring into your brain. You have to deal with it along with the important task you're supposed to be fully focused on. Because you're doing two things at once, you fail at your main task.

The third theory suggests that overarousal and overstimulation, especially in extreme situations—such as competing for an Olympic medal—make people unable to perform at peak levels. Excess energy has nowhere to go, according to this theory, and the result is a lack of focus.

Sian Beilock at the University of Chicago has tried to make sense of the third theory by actually studying putters. She set out to see if putters were most likely to choke when they allowed their conscious mind to intrude at the very worst time. She worked with students, and although they weren't exactly pros, as part of the experiment they hit hundreds of practice putts, enough to make them sufficiently competitive. The students were then put into three groups. The first putted knowing there was money at stake; the second while forced to pay attention to distracting words; and the third in front of a video camera.

The group with money at stake and the distracted group each putted worse when faced with these challenges compared to putting without them. But the students putting while being videotaped did fine—their performance didn't decline with that challenge in place. That was interpreted to mean that these students were used to being recorded, and so they didn't have to think about being videotaped and avoided having it affect their performance. They were, in effect, inoculated against the pressure that leads to choking, whereas the others weren't.

One thing Beilock found in other studies was that expert players can detail for you the important parts of a good putting stroke, but they can't tell you nearly as much about the last putt they made, whereas novices are exactly the opposite. Psychologists call this expertise-induced amnesia. For experts, their skill is automatic, run by neural networks they're not even aware of. It's the difference between learning to drive and being an expert—after a while you don't think about the individual steps, you just do them.

You have "expertise-induced amnesia." I recommend forgetting everything you know.

A different study, which also observed that consciousness can induce performance failure, showed that if you count backward from one hundred while you're executing a skill, like putting, you can occupy your conscious mind. It's too busy counting to mess you up in other ways, and so you survive the pressure and succeed.

It's also true that the process of becoming an expert in any activity is likely to change the physical brain. A nice example is juggling, an activity where thinking about what you're doing could be fatal if you're juggling firesticks or chainsaws! With training, juggling becomes an unconscious routine, and at the same time, new neurons and connections to them are added to the motor control parts of the brain. Stop the training and the brain returns to its former state. But even with these additional brain cells, if unusual pressure is put on the juggler, other areas of the brain interfere, with unfortunate results.

Most of the thinking about choking involves differentiating between the conscious and the unconscious. Add up all the information flowing into your brain from your five senses, and the estimates are that maybe one-millionth of that information actually enters awareness. All the rest is unconscious. If we return to the example of driving, you'll see ample evidence of this: you don't think about the details of driving ("now check the rearview mirror, now look up ahead, now check my speed, now apply more pressure to the gas pedal"), you just drive. Your unconscious mind takes care of the subtle operations. Now put yourself on the putting green, with the distraction of cheering crowds, the clicks of cameras and $100,000 at stake. Pretty hard not to overthink your putting stroke. But if you do . . . choke!

When the pressure is on, I get this lump in my throat...

Me too.

Why do farts smell bad?

EXPERIMENTS TO EVALUATE FARTS aren't all that easy to perform. It doesn't take much to persuade volunteers to be farters, but it takes a lot to persuade them to be smellers. The few experiments that have been run, though, have all found the same thing: where there's stink, there's sulfur.

In one infamous experiment, ten men and six women were put to work eating pinto beans and lactulose (a synthetic sugar usually used to treat constipation). The volunteers were then instructed to fart into a number of test tubes and then quickly seal them up to trap the gas. The researchers analyzed the gas to see if it contained any sulfur compounds. They weren't surprised when they found a cocktail of sulfurous chemicals.

Then came the hard part. The scientists gave the tubes to a group of judges and asked them to rank the intensity of the smell in each one. When the researchers compared the judges' results with their chemical analysis, they found that the gases that ranked highest on the judges' intensity scale also had the highest sulfur content.

I've been a volunteer fart-sniffer for years. I'll quit when it stops being a gas.

Science _Fiction!_ *The idea that men fart more than women might sound correct, but there's no truth to the claim. In fact, most people release about 1 quart (1 liter) of farts each day, and on average, women's flatulence contains higher concentrations of hydrogen sulfide, the smelliest chemical in farts.*

Most people would have stopped the research at this point. But the scientists weren't happy with just establishing that sulfur was the leading ingredient in the stink. They went one step further: they decided to test different materials to try to find one that would capture and hold on to the chemicals in the fart odor.

This had to be a much more tightly controlled experiment than the one with the tubes of farts. The researchers outfitted volunteers in pants specially designed with a lining made of either activated charcoal or zinc acetate (a salt that's often found in cold medications). In this experiment, the volunteers didn't have to produce their own farts—there's too much variation from person to person. Instead the scientists concocted their own uniform fart-like mix of gases and pumped it into the airspace near the anus of each volunteer. A short time later the scientists recaptured the gas and compared it to the original concoction to see how much of the sulfur—and thus the stink—had been captured by the special pants. The charcoal pants performed the best, removing almost all the gas and odor. The salt pants removed most of the gas, but didn't successfully eliminate the stench.

Did You Know . . . Although farts stink, your breath is much worse. Somewhere between one hundred and two hundred chemicals conspire to make your breath smell bad. Within that mix, the most potent ones contain—you guessed it—sulfur.

The amount of work that went into researching these specially designed pants might seem over the top, but it's useful if you imagine a close-quarters cross-country airplane trip. Flatulence can be particularly difficult in an airplane. Any gas that's inside you at takeoff will increase in volume by about a third by the time you reach cruising altitude. That thought should persuade you to release any gas you have before you take off (and preferably before you board the plane), but if you miss that opportunity, medical professionals agree it's far better to let it rip than to hold in gas during a flight.

Letting loose like that might make you more comfortable, but it can lead to nightmares for your seat neighbors. After all, when all that sulfur comes out, there's nowhere it can really go in an airplane. Some airlines already use activated charcoal in their air conditioning to keep the cabin air fresh, while others have looked at installing charcoal-infused cushions to absorb the smell. The problem with the cushions is that, if a person is wearing pants or a skirt made of leather or some other material that won't let the fart pass through it (it's called "low fart permeability"), then their clothes could create a tunnel effect. The fart might be channeled down the pant leg and out into the open air, completely bypassing the charcoal-lined cushion. Researchers also point out that business-class leather seats wouldn't permit gases to seep into the charcoal cushion either.

It seems, then, that for frequent fliers, a pair of charcoal-lined underwear might allow them to sneak a fart while still keeping their fellow passengers happy. Of course, that strategy would be completely undermined if they forgot to wear underwear at all.

Why are yawns contagious?

ABOUT 50 PERCENT OF US ARE SUSCEPTIBLE TO CONTAGIOUS YAWNS: that means we feel the irrepressible urge to yawn as soon as we're aware that someone else is yawning. I say "aware" because you don't have to actually see a yawn in order for it to be transmissible. A contagious yawn can be triggered by hearing a yawn, overhearing a person talking about a yawn or even just reading about one. The yawner could be in a video you're watching, and the video could be sideways or upside down. If you're one of these people who falls prey to the contagious yawn, you know that it is never really your choice to yawn; you can't stop it. But why do we share this behavior? What's the point?

Robert Provine at the University of Maryland is one of the world's most prominent yawning researchers, and he has established the basics of the process. On average, an ordinary yawn lasts six seconds and involves a huge inhalation followed by an exhalation, stretching the

mouth open to its limits and squinting the eyes. There's a repeatable series of events in a yawn—eyes close, mouth opens, air moves in and then out, yawner relaxes. Once you start the sequence, it's hard to stop it, and if you do, you feel unfulfilled. You can yawn while pinching your nose closed, as odd as it feels. But try it with your teeth clenched and you'll find a yawn is very difficult, if not impossible, to complete. That's curious, because if the purpose of a yawn is to move a lot of air in and out, you can do that perfectly well with your teeth clenched. Conversely, you can have a huge intake and output of air and still have a failed yawn. You need the whole package of facial actions to propel your yawn to completion.

Provine has tried to identify exactly what parts of the yawning face trigger the contagion of a yawn. Surprisingly, the mouth isn't that feature: yawners whose mouths are obscured nonetheless prompt yawns in those watching them. The flip side is that a yawning mouth on its own isn't easily recognized as yawning—it could just as well be yelling. Also, if you cover the mouth during a yawn and just watch the other parts of the face, then a yawn looks a lot like an orgasm. The two have similar dynamics—the buildup to a climactic moment, followed by the return to a baseline—suggesting they might be sharing fundamentally similar low-level brain mechanisms, ones that have been around for millions of years. No one has yet addressed whether you can yawn and have an orgasm at the same time. Maybe you can't because there's only one set of facial muscles available for both?

More than just tracking the mechanics of a yawn, though, Provine has disproved the common theory that yawning is a response to inadequate oxygen or too much carbon dioxide in your blood. He's shown beyond doubt that neither reducing oxygen nor increasing carbon dioxide increases the frequency of yawning. According to Provine, the epidemic of yawns triggered by a stuffy lecture hall in the late afternoon has less to do with the air in the room and more to do with the lecture itself. Provine tested this idea. He found that students watching static on a screen yawn more often than their counterparts watching videos. It was boredom and fatigue, then, that led to increased yawning.

That makes sense in the late-afternoon lecture hall. But it's hard to see why boredom, which comes in many forms, would instigate a behavioral response as universal as a yawn. Restlessness, yes. Going to the fridge, maybe. But opening your mouth wide and closing it? There's no obvious connection.

Did You Know . . . Scientists in China and at the University of Michigan are developing software that can analyze images from a dashboard camera to detect driver fatigue. Yawning is a reliable clue to fatigue, but capturing a moving driver's face in variable light conditions isn't easy. The system is designed to detect the indicators of yawning while ignoring the mouth. The setup so far includes a face detector, a nose detector, a nose tracker (to follow the nose's movements) and a yawn detector. Scientists working on a similar detector at the University of Strathclyde are also concentrating on the rest of the yawning driver's face rather than the mouth, on the sensible grounds that people do often cover their mouths when they yawn.

Another theory is that yawns cool the brain. This could be useful in warm weather—and there is evidence that people yawn more often in warm temperatures than cold—or when the brain literally heats up if you've been thinking hard. Rats with mini-thermometers in their brains exhibit a quick rise in temperature immediately before a yawn, although this effect has been questioned because the rats are yawning and stretching simultaneously, so it's hard to know which causes the spike in temperature. It's also not clear that the cooling effect lasts for any appreciable time.

In the end, it might turn out that contagious yawns have both physiological and psychological causes. Some psychological explanations trace yawn contagiousness all the way back to early hominids. According to those theories, yawning was supposed to be a signal passed around the group, to heighten awareness, to start moving together or to bed down for the night. There's evidence that passing yawns around a group establishes empathy. That's a tricky proposition to test directly, but there are some clues. Children don't generally participate in contagious yawning until they're four or five years old, when they acquire something called theory of mind, which is essentially the ability to understand that others are having thoughts and emotions of their own. Children have been yawning spontaneously since they were in the womb, though, suggesting there are two different mechanisms in play—one for a standard yawn and another for a contagious one.

Acquiring theory of mind is a crucial development for social beings like us. Some children with autism spectrum disorder are slower to (or never) achieve theory of mind, and that affects the development of their social lives. They are not susceptible to contagious yawning. Neither are people who, while not full-on psychopaths, have some psychopathic traits, like selfishness, manipulation, impulsiveness, callousness, dominance or, above all, lack of empathy. The more psychopathic traits an individual displays, the less likely he or she is to yawn in response to someone else's yawning. One 2016 study found that females are more susceptible to contagious yawns, a result that was immediately recruited to support the "empathy" idea.

Other explanations for why yawns are contagious take a wildly different approach, looking at how yawns occur in other species. Chimpanzees frequently respond to a yawn with one of their own, but they do so only in response to yawns from other chimps, not humans. High-status baboons yawn a lot, and their yawns have the happy coincidence of exposing their giant, threatening canine teeth. Baboons on the lower social rungs are usually wise enough not to yawn back, as contagiousness in those situations might be fatal. It was once suggested that this threat avoidance was the origin of covering your mouth when you yawn. The problem is that, even if you did cover your mouth, the aggressor might still know you were yawning just by looking at the rest of your facial expression.

Monkey see,
monkey do.

A 2008 study stirred a lot of excitement by claiming that dogs were prompted to yawn when they saw other dogs, or people, yawning, and that this reaction occurred at a higher rate than it did in humans. This was a major coup for the empathy idea, given that dogs have made it to where they are today by exploiting empathy. Unfortunately, other studies have failed to confirm the original findings, so we're left with a possibility that dogs respond to yawns, too, but really no definitive evidence to back up the claim.

Did You Know . . . Budgies yawn, but they're the wild card of contagious yawning. When budgies are allowed to perch across from each other, separated only by glass, their yawns cluster together in time. The response yawns are not rapid-fire by any means: some of them take minutes to appear. But that's still significantly different from when the budgies can't see each other; in that case, any yawns are not connected time-wise at all. Birds are a lot more intelligent than we have given them credit for, but an empathetic budgie really stretches the list of animals known to experience contagious yawning: humans, chimps, dogs and high-frequency-yawning Sprague-Dawley rats.

There has been some mapping of what goes on in the human brain during a yawn, and one revealing discovery was the absence of activity of mirror neurons. Mirror neurons are brain cells dedicated to monitoring the behavior of others and imitating it. That they are not involved in yawning suggests that the signals kicking off the yawn are coming from somewhere deeper in the brain. This would mean, first, that it's more or less automatic and, second, that it likely goes far back into our ancestry. It is a deep-seated behavior.

Stifling a contagious yawn is different—you have to think to do it. Cells in the prefrontal cortex, which are involved in thoughts and actions involving empathy, are active during contagious yawning. The prefrontal cortex is silent during spontaneous yawns, though, suggesting a different neural circuitry is involved in the two kinds of yawning. How remarkable it would be if it turned out that over the course of our evolution we co-opted an established physiological reflex for shaking ourselves out of fatigue and turned it into an empathy signal?

History Mystery

*Did Newton really get hit on the head by an apple,
inspiring his thoughts of gravity?*

THIS IS THE BEST-KNOWN SCIENCE STORY EVER: Newton, sitting in the
garden, musing under an apple tree. An apple falls from the tree and inspires
him to invent a whole new way of thinking about gravity.

Did it really happen? Unfortunately, Newton himself was mostly mum on the subject. He never said anything about this moment (which supposedly happened when he was in his early twenties) until he was nearing death, at the age of eighty-four. At that time, sitting in a different garden with his old friend William Stukeley, he explained how he'd begun to question gravity based on the way an apple falls: Why always straight down? If it were just Stukeley's claim to have heard Newton say this, it might have been forgotten, but three other writers claimed Newton told them the same story. And as with all stories, there are inconsistencies—for instance, one person mentions the garden but not the apple.

Science Fiction! *While Newton may very well have been sitting under an apple tree when a piece of fruit fell, inspiring thoughts about gravity, he was never hit on the head by that apple. That part seems to be totally fictitious. But it does make the story more colorful!*

Newton wouldn't know gravity if it hit him on the head... which I didn't!

The notion of gravity strikes us now as common sense, but in Newton's time, it wasn't. In the 1660s, Newton had returned home from Cambridge University because of the threat of the Black Death. He was thinking hard about gravity, a force that to most people was mysterious. There were only vague and somewhat mystical explanations about why things moved the way they did, such as the idea that invisible strings were attached to everything, and heavier things had more strings. Or the theory that while some things obeyed gravity and fell to earth, others, such as smoke, were subject to "levity" and rose instead of falling.

The grand ideas of attraction and movement were on Newton's mind, and so it isn't unreasonable to think that the sight of an apple falling to the ground might have caught his eye. But that's not the amazing part. What's amazing is what he did with the idea. He told Stukeley that it occurred to him that the earth draws the apple toward it, but the apple must also draw the earth. He also linked the fall of the apple to the movement of the moon in the sky, concluding that they were both examples of the same force. The moon was, in effect, falling toward the earth, but it never hit because its speed carried it around the earth in an endless lunar orbit. This thinking, rendered in complex math and extended to the visible universe, was the basis for Newton's mighty *Principia*, in which he laid out his famous laws of motion in 1687.

Just to put this in an even larger and more impressive context: Newton was either twenty-three or twenty-four when that apple fell. He was thinking about gravity but also busy inventing calculus and figuring out the nature of light (determining that light was composed of the colors of the rainbow). Absolutely mind-boggling.

Not all Newton scholars love the story of the apple. One has written that it is "vulgar" to think that the mere fall of an apple inspired the mighty laws of gravitation. But the apple wasn't all there was to it: it was another twenty years before Newton published all the complex ideas that resulted from his thoughts about gravity. Also, what's the problem with a commonplace thing being inspirational? That inspiration still requires a great mind. In this case, one of the greatest ever.

Science _Fact!_ Newton's apple tree, at Woolsthorpe Manor in England, has a colorful history. It was the only apple tree in the garden at the time that Newton rested there, and a century and a half after its most famous apple fell, it was blown down in a storm. It managed to survive long enough to procreate, though. Cuttings from the tree and its descendents have been taken many, many times and grafted onto apple trees all over the world. In Canada, there are two descendants of Newton's tree—one at the University of British Columbia, the other at York University in Toronto.

There is still a mystery, though, because genetic analyses reveal that two trees have given rise to all of Newton's trees around the world. That means that some proud owners of this vestige-of-a-great-moment-in-science may have nothing more than a regular apple tree. The trees do produce apples, though—a variety of cooking apple called Flower of Kent, said to be good for cooking, not eating, because it is mealy and flavorless.

Part 2
The Animal Kingdom

Where do cats come from?

Cats, like dogs, have ancient origins, but the route they took to get where they are today was very different from that of their canine nemeses.

Ten-thousand-year-old cat bones were found buried alongside humans on the island of Cyprus. Those cats didn't swim there, so they must have been taken there by humans. And given their esteemed burial spot, they were clearly pets. It's assumed that domesticated cats date back to a few hundred, or even a thousand, years earlier than that, making their debut around eleven thousand years ago.

*Science **Fiction!*** *Although many believe the house cat comes from Egypt, the Middle East is more geographically accurate as the home of the first domesticated kitties. Egyptians definitely worshipped the feline, though, elevating them to god status. Cat mummies abound, and archeological digs have even uncovered cat cemeteries.*

Genome analysis has identified the wildcat as the likely ancestor of the house cat—and one species in particular. The domestic cat lines up genetically with the Middle Eastern version of the wildcat, considered the least fearful and most gentle of all of its kind, and therefore the most likely to be domesticated.

There is another factor that leads us to conclude the Middle East is the likely geographical origin of the house cat—the rise of agriculture. Ten thousand years ago or so, our ancestors abandoned nomadic life and began to settle on the land, growing and storing their food. The downside of storing grain is that you are forced to share it with rats and mice. Small rodents are the perfect diet for wildcats, and so, at least the theory goes, granaries attracted mice, which in turn attracted cats to agricultural settlements. Perhaps cats and humans didn't interact much at first, but there would have to have been some accommodation on both sides, and why not? There were obvious benefits to both. It's also quite likely that at some point, kittens from the granary were adopted as house pets. These lucky cats probably ate scraps from the dining table and forged closer bonds with people.

Did You Know . . . Black cats are everywhere at Halloween, for good reason. In the Middle Ages, the Catholic Church sought to eliminate heretical pagan groups such as the Knights Templar. To pagans, cats represented fertility, but Christians associated cat-worship with the devil, cannibalism, orgies, and the execution of children. This association lasted for centuries, and it's a testament to the species that cats survived, despite the incredible persecution they endured.

Examination of the cat genome reveals the changes wrought since felines moved in with us, although these adaptations are not as widespread as those in dogs (because cats are more recently domesticated). Cats continue to retain many of their wild attributes: the broadest hearing range of any carnivore (into the ultrasonic to hear those little mouse conversations), fantastic night vision and adaptation to a fully carnivorous diet. However, the cat genome shows clear changes in areas of the brain that regulate timidity, fear and reward-seeking. Overcoming timidity and learning to beg for rewards would have been helpful in ensuring their place with us.

Did You Know . . . Cats have lost the supreme sense of smell that dogs have, which is understandable because they are primarily visual hunters. At the same time, they have amped up their nasal detection of pheromones, the subtle chemical signals that regulate social behavior.

Having originated in the Middle East, cats from that founding population were taken around the world by humans. The riot of coat colors that domestic cats now exhibit was bred into them by us over the last two centuries—a blink of an eye in the evolution of an animal. Is it possible that over time cats will become even more domesticated? Who knows. Perhaps one day, they'll all roll over onto their backs and let us rub their bellies the way dogs do. Or maybe not.

Where do dogs come from?

DOGS CAME FROM WOLVES. That seems straightforward, but if you look at a Chihuahua, a Great Dane, and a wolf side by side, you'd be right to wonder if there's more to the story of dogs' ancestry. Knowing that dogs came from wolves is just the start; it's why, how and when that happened that are the really interesting questions.

Hey, pops!

The first attempts to figure out where dogs came from relied on archeology. Researchers have uncovered skulls of dog-like animals that date back to more than thirty thousand years ago. The skulls they found were a long way from anything we'd recognize today, though. At least in the very early transition from wolf to dog, several physical features of the animal head's changed: the teeth became more crowded and the skull grew wider. To put this into perspective, if modern dogs were on the scene more than thirty thousand years ago, that means they were in Europe shortly after the Neanderthals died out. That great depth of time sheds some light on how the wolves of the time might have become domesticated.

There used to be a popular idea that one day, an enterprising ancestor of ours came upon a den of newborn wolves, scooped one up and tamed it, giving rise to the millions of dogs in the world today. But the reality is that wolves played as much of a role in their taming as we did.

Stay. Sit. Roll over.

Thirty thousand years ago, our ancestors were still nomadic hunter-gatherers, constantly relocating as they searched for prey such as deer, smaller mammals and edible plants. The wolves of that time—and most researchers now agree that they were not the same animal as the modern wolf *Canis lupus*—were also on the move, looking for prey that they would attack and devour as a pack. Packs of wolves and humans on the hunt would undoubtedly have come into contact. Humans wouldn't have looked on wolves as prey and nor would the wolves have seen humans that way. But their proximity means that wolves would have encountered carcasses left behind by humans. Smart animals are able to put two and two together: humans (inadvertently) provide food. So, the theory goes, packs of wolves would have started to drift along with human hunting parties to pick up their leftovers. And the humans would have benefitted from keeping the wolves close—packs of the animals living on the edge of their hunting camps would have alerted the humans to the presence of either predators or prey nearby.

 **Did You Know . . .** There was a huge population crash among the wolves that gave rise to dogs. It happened about fifteen thousand years ago—the same time as the conclusion of the last ice age—as a result of an unknown environmental incident. The crash might explain why no one has found any remains of the wolf ancestor of modern dogs.

These archeological theories were useful up to a point, but they didn't explain exactly how the wild wolves of the past turned into the dogs we know today. To solve that problem, scientists had to turn to genetics. One of the earliest experiments to use this new approach was run in the late 1950s by Dimitri Belyaev in the Soviet Union. Belyaev believed that the tamer an animal was—or the more easily it got along with humans—the easier it would be to domesticate. To test his idea, he began a program of breeding silver foxes based on how tame they were. He started the program in the 1950s and worked on it until he died twenty-six years later. By evaluating how eager a young fox was to accept a human's approach or how willing it was to hang out with a human and not bite them, Belyaev could establish how tame that animal was. After examining each generation of foxes, Belyaev would allow only the top 20 percent of them—the ones he found to be the least fearful and aggressive—to breed.

As the years went by, the foxes grew increasingly tamer with each generation, but they also changed physically and behaviorally: they started whimpering to attract human attention, licking their trainer's hands and wagging their tails. Their heads changed shape, their ears became floppy instead of standing up straight, their tails started to get curly and their fur changed color. Even their hormones changed from those of their wilder ancestors.

Belyaev's experiments suggested to scientists that a similar kind of selection might have happened around our ancestors' campfires. The less aggressive and less fearful wolves would have established themselves as part of the human camp and bred with one another, making each new generation tamer and changing their physical features until they became the dogs we know today. Recent work by scientists hasn't been able to determine exactly when this change happened. One study found that wolves finished their evolution into the modern dog around eleven thousand to sixteen thousand years ago. Another study sequenced the genomes of fifty wolves and dogs from around the world and concluded that the domestic dog originated in Southeast Asia around thirty-three thousand years ago.

*Science **Fiction!** Some theories have claimed that dogs didn't appear until humans began engaging in agriculture. Scientists used to think that dogs' genetic code had a starch-processing enzyme, amylase, that wolves lacked. Because starch would be found only in food or waste left over from farmers, not hunter-gatherers, the scientists reasoned that dogs could only have developed after humans gave up their hunter-gatherer ways to take up farming. That idea was put to rest when recent genetic tests found that wolves do indeed have the amylase gene.*

In the course of dogs' evolution, they developed one of their most distinguishing characteristics: their bark. Fully grown wolves hardly ever bark, but wolf pups do. Somehow, as dogs became domesticated, they retained that juvenile characteristic as a part of their adult lives. Not only that, but dogs refined barking into a useful communication tool.

Sophia Yin and Brenda McGowan at the University of California recorded a grand total of 4,672 barks uttered by a mix of ten breeds. They recorded the dogs' barks in three different situations: when a stranger rang the doorbell, when a dog was separated from its owner by a locked door and when the dog was playing with either a human or another dog.

When they analyzed the barks acoustically, they found that there was a wide variety ranging from harsh low-frequency ones to higher-frequency barks rich in harmonics. The barks directed at the stranger ringing the doorbell were harsh and long in duration, but the barks while the dogs were separated or playing were higher frequency with long spaces between them. There is a general rule in the animal world that a harsh sound means hostility, while higher pitched, musical sounds are more social. By that measure, dogs appear to be wilder than we thought.

Wherever and whenever dogs originated, it's obvious that they eventually migrated all over the world, either on their own or with humans. But although it's been a very long time since the wolf and the dog separated, the two species can mate and have fertile offspring even now, so genetically they are still extremely similar. So similar, in fact, that their mitochondrial DNA—the genes that are passed down strictly through the maternal line—is 99.9 percent the same. Half of that 0.1 percent difference, though, is linked to the brain and social interactions. So you can thank that tiny fraction the next time Fido licks your hand instead of biting it.

Do cats love us
or merely tolerate us?

Dog behavior is well studied and understood, whereas cat behavior is still a bit of a mystery. Roughly speaking, the difference can be summed up like this: dogs think we are the lead dog and act accordingly, whereas cats believe we are large, dumb versions of their mothers.

Man's best
friend.

Man's best
fiend.

When your cat kills a mouse, she brings it to you and drops it at your feet the same way she would for her kittens. Desmond Morris, the zoologist who gained fame by writing *The Naked Ape*, argues that the female cat leaving you prey is repeating mother—kitten behavior, intending that as "deputy kitten" you accept her gift gracefully—and, presumably, eat it. But, male cats do not bring mice to kittens–but do bring mice to their owners—which puts that whole hypothesis in doubt.

There is disagreement on other aspects of cat behavior as well, even around the litter box. The one thing you can be pretty sure of is that cats are not using the litter box to please us. House cats generally cover their feces in the litter box, with the common explanations being that it hides their presence from nearby predators or prevents the spread of parasites. Outdoor cats, however, behave differently. In one study, a researcher followed female house cats doing their business outside. She watched them defecate fifty-eight times, but only twice did a cat actually dig a hole for the waste and only about half of the fecal deposits were partially covered. And the farther a cat was from home, the less care she took to cover her feces. It could be that cats note some transition on the border between home territory and the rest of the world. Other than that, we just don't know.

What about when cats press forward repeatedly with their front paws as if they are kneading dough? In theory, they're doing what they did to prompt their mother to allow them to nurse. But surely they don't expect us to nurse them. Or do they? When they rub against us with raised tails, they're showing signs of friendliness—just as they would with other cats.

Did You Know . . . Cats are much more vocal with human beings than they are with other cats. In feral cat colonies, meowing is very rare. Yet house cats have an array of meows, moans, howls and growls that seem to help them communicate with us. There is evidence that people can rate the sounds of cats in terms of their pleasantness or urgency, suggesting cats use sound to stimulate appropriate behaviors in their owners.

Universal Cat Translator

Meow.

You annoy me.

Mee-ow.

Feed me. Now.

Mrreow.

I defecated in your bed.
You're welcome.

Meeew.

If you touch my belly one
more time, I will destroy you.

Purring is another confounding aspect of cat behavior. The unusual sound is created by rubbing the vocal cords together, rather than forcing air past them. It is generally thought to be a noise made either to signify security or to seek comfort, because cats purr not only when they're curled up beside you but also when they're afraid or in pain.

Karen McComb, a specialist in animal behavior at the University of Sussex, noticed that her cat Pepo woke her up in the mornings with an unusual purr that indicated he wanted to be fed. She then discovered that several other cat owners had experienced this particular food-seeking purr. It was unusual because it sounded different, and when McComb recorded these novel purrs, she discovered they included a tone not present in regular purring. She had both cat owners and others listen to these purrs and rate them. The tone made the purrs sound more urgent and "unpleasant" to most listeners (and therefore more likely to be acted upon, if for no other reason than to shut the cat up). What was truly striking was that this added frequency was right in the middle of the range of human infants' cries, suggesting cats were embellishing their purring to make people pay attention.

It's clear that we're still in the dark when it comes to understanding all the signals cats send our way. This gives cat lovers leeway to make the case that cats, unlike dogs, simply don't care about pleasing us and are therefore superior. Thousands of years ago, cats were worshipped as gods. It's been said they've never forgotten that.

Are dogs truly man's best friend?

Dogs DESCEND FROM WOLVES, and they bring that history to their relationships with us. Because wolves are pack hunters, social organization is of the utmost importance, and so they have a highly developed set of signals they use to communicate with each other. Their cues include facial expressions like baring their teeth or laying back their ears, body language like keeping their tail erect or dropping it between their legs, and, of course, sounds such as barks and howls. Wolves also have astounding olfactory capabilities, and they use this remarkable ability to gain information about other wolves from the scent left behind in their urine. Wolf packs are typically composed of extended families organized around a leader (the animal traditionally called the alpha male), and so the communication that young wolves learn runs to them from the head of their group. In the long evolution away from wolves, though, dogs have learned to adapt these abilities and interpret messages from a very different leader: humans.

Did You Know . . . Wolves' and dogs' noses are one hundred million times more sensitive than ours.

Dogs relate to their humans in much the same way as human babies do to their mothers, and dogs are much more attuned to human gestures than other animals. They naturally find their place in our social hierarchy by carefully tuning in to all the cues our voices and gestures provide. One experiment compared wolves, chimpanzees and dogs in their ability to pick up on human signals. The experimenters set out several crates, only one of which had food in it. They then brought one of the animals into the room, and the human either looked at, pointed to or tapped the container with the food in it. Dogs were quick to figure out what the gesture meant, but the wolves and chimps never learned it.

The researchers then gave the experiment a twist. To ensure that the wolves were socialized to humans, they had a litter of pups live with humans for twenty-four hours a day from the time they were four days old. When the wolves were grown, they were as used to humans' voices and gestures as were the dogs. When the scientists re-ran the experiment, a few of the wolves were able to follow some of the cues, but overall they still weren't as good as the dogs. It was clear that the dogs had a deep understanding of humans (even though on average their brains are 10 percent smaller than wolves').

Little help?

A later experiment finally showed the crucial difference between the two animals. Scientists trained a group of dogs and wolves to retrieve a piece of meat in two ways: either by lifting the lid of a bin to get at the meat inside or by pulling a rope (tied around a piece of meat) out of a cage. Once the animals had figured out how to do both, the scientists tricked them: the lid of the bin was locked down and the rope was tied to the cage. The tasks were now impossible. The animals had very different responses to the problem. Where the wolves focused on the challenge, the dogs kept looking at the humans present, apparently waiting for some sort of clue to help them get the meat. Seven of the nine dogs kept looking to the human, while only two of seven wolves did. The scientists concluded that the main difference between the two animals was that dogs scrutinized the faces of their humans.

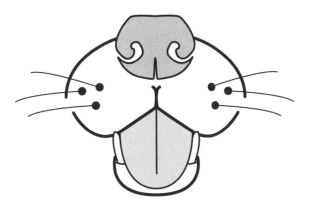

Dogs' abilities to read our faces have likely been adapted from their behavior with other dogs, where they read emotional signals and copy them, something called emotional contagion. Emotional contagion is a way of establishing empathy. An Italian team watched a pack of dogs playing in a dog park in Palermo, Italy. They saw that when two dogs met, one of them would copy the expression and behavior of the other, often in less than a second. It was out-and-out mimicry meant to strengthen the social bond between the two animals, and the researchers suspected that the same thing goes on between dogs and their owners.

Science Fiction! *Some people think that dogs know when their owner is coming home. One case study involved a dog named Jaytee and his owner, Pam. One video camera stayed with the dog at home while the other followed Pam into town. At a time known only to Pam and the researchers, she got into a taxi and returned home. Eleven seconds after she got into the cab, the camera at home shows Jaytee getting up and going to the porch. It looks perfect, but there are unanswered questions. For example, skeptics like Richard Wiseman have suggested that Jaytee might have become more anxious the longer Pam was away and visited the porch more often for that reason. No experiments that satisfy everyone have been done yet. Most scientists can't accept that the dog would have some special, long-distance knowledge of his owner's movements, an ability that would hint strongly of telepathy. That tends to make most scientists uneasy.*

Although dogs are uniquely tuned in to our body signals, it is sometimes tempting to read too much into their behavior. My favorite experiment of this kind was one that tested dogs' tendency to look guilty whenever they do something wrong. In the study, owners ordered their dogs not to eat any treats, then left the room. When they were outside, the researchers either gave the dog a forbidden treat or did nothing, then invited the owners back in. Some owners were told that their dog hadn't eaten the treat (when the dog actually had), while others were informed that their dog had disobeyed the command and eaten the treat (when it actually had not). The owners were encouraged to either pet or scold their dog, depending on what they'd been told it had done.

If an owner chastised a dog—whether it had eaten the treat or not—the animal looked guilty, slinking low to the ground and avoiding eye contact. In fact, the dogs that had obeyed the order were judged to look the guiltiest when disciplined by their misinformed owner, suggesting that a dogs' guilty look wasn't in response to an actual feeling of remorse—it was just a reaction to its owner's behavior. Remember that next time you go to scold your dog for eating the leftover pizza: Fido is reacting to your voice, but he doesn't regret the pepperoni one bit.

Mmmmm

What attracts mosquitoes to me? (And what can I do about it?)

MOSQUITOES ARE MORE ANNOYING THAN DANGEROUS for most of us in North America. I don't mean to trivialize West Nile virus, nor am I ignoring the fact that with climate change, more mosquito-borne diseases are beginning to creep into the southern United States. But for most of us, it's those buggy summer nights that make us wish them extinct. That's not likely to happen any time soon, so in the meantime we search for ways to make ourselves unattractive or, even better, invisible to our biting nemeses. There's just one problem: mosquitoes are finely tuned for searching us out—so much so that after reading this, you might lose faith that you will ever find peace without taking refuge indoors. And even then . . .

Mosquitoes have a package of detectors and responses that would challenge the most powerful computers linked to the most sophisticated robots, all operated by a brain that is about the size of the period at the end of this sentence.

This is the size of a mosquito brain

These detectors include human-targeting sensors that are turned on in sequence as the mosquito closes in on a person.

Imagine you're sitting outside at dinnertime in the summer in an area frequented by mosquitoes. Assume there's not too much wind, no campfire—just you. You are breathing—that is, exhaling carbon dioxide. As that plume of gas wafts through the evening air and breaks up, a wisp of it passes over a mosquito's head, coming into contact with specialized CO_2 receptors on her maxillary palps (antenna-like extensions on her head). A volley of nerve impulses is triggered in the bug's brain. She's on your trail.

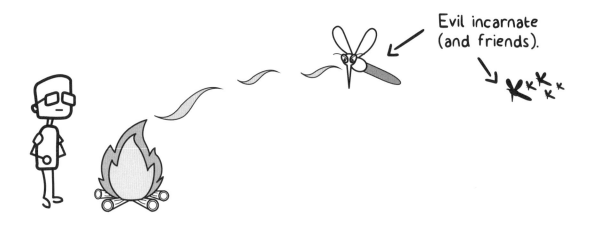

Evil incarnate (and friends).

Now she moves toward the source of the carbon dioxide—you. Not only has she detected the gas, she's adjusting her overall flight direction toward higher concentrations—all of this even though carbon dioxide sensors on the palps represent only 2 percent of her total sensory array. But it's not enough for her to depend on finding you by CO_2 alone—otherwise she would fly directly into your mouth, and while that would be annoying to you, it wouldn't help her at all. At some point she needs to switch her attention to other attractants to be able to close in on your bare skin.

I'm a real mother sucker.

Did You Know . . . Only the female mosquito bites. She needs the protein in your blood to develop her eggs.

In a beautiful set of experiments, Caltech researchers showed that once a mosquito detects carbon dioxide, she then pays attention to visual features in her environment. To prove this, researchers had female mosquitoes fly freely in a wind tunnel that was visually plain except for a single dark square on the floor at the end (mosquitoes' visual systems are more sensitive to dark than light). In the presence of normal air, mosquitoes paid no attention to the floor at all, but when the experimenters introduced carbon dioxide into the tunnel, the mosquitoes immediately began to inspect the dark square on the floor, spending hours hovering about an inch above it. They had switched over from gas detection to vision, trying to home in on a potential target.

Science Fact! *It's true that wearing light-colored, high-contrast clothing will help you stay invisible to mosquitoes, but it messes up only one of their many detection systems. Scientists agree that wearing light clothing alone is not enough to thwart the "annoyingly robust" sensory powers of the female mosquito.*

In the Caltech experiments, once the mosquitoes closed in on a target using vision, they then deployed heat- and chemical-sensing to find the right landing spot. At that stage, warm objects attracted mosquitoes much more strongly than ones at room temperature. Finally, mosquitoes were attracted even more strongly to warm things with high humidity—like your sweaty skin.

And just in case you're thinking there might be a way of circumventing this string of sensory modules that turns you into guaranteed mosquito bait, think again. As it turns out, a mosquito can even skip one of her detection tools and still find you without much trouble. Let's say you're holding your breath so she doesn't detect your CO_2; she can still be attracted to your body heat as long as she flies close enough to you. And all she really needs is someone close to you to exhale and she'll be on her way over to you and your breathing buddy. If you could persuade your friends to stop breathing alongside you, that might help you stay incognito—but not entirely. If you're warm and sweaty and happen to be wearing dark clothing, you're still giving yourself away.

Why am I plagued by mosquitoes when some of my friends aren't?

YOU ARE NOT MAD. It's true: some people are particularly attractive to mosquitoes. I would suggest befriending them to avoid detection yourself. Other than that, I'm sorry to say there's no definitive answer as to why mosquitoes prefer some people over others. But there are a few interesting leads.

People are great. They always clap when I come around.

One stunning experiment executed decades ago in California tested (and that is definitely the right word) 838 volunteers by exposing their bare arms to mosquitoes for three minutes. In the first round, only seventeen of the volunteers weren't bitten. In the next round, only one remained unscathed. That guy was tested nine more times, and never, ever was he bitten! Unfortunately, what set him apart from the rest of the human race was never identified. Some theories suggest that it could be chemicals in the breath, although I'd bet if you had people sitting on a back deck wearing mouthpieces attached to giant air storage tanks, you'd still see differences in how attractive they are to mosquitoes. (That experiment hasn't been done yet. Any takers?)

Mosquito Taste Test

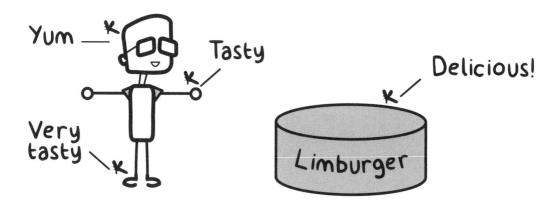

It's also been said that beer consumption increases your mosquito allure. For the good of humankind, I participated in a beer–mosquito experiment at Brock University in St. Catharines, Ontario. While I can say that it afforded me a pretty good time (the beer part, not the bite part), the results revealed that my intoxication levels had little to no effect on mosquitoes' interest in me.

Did You Know . . . Feet, hands and face—in that order—are the preferential feeding sites for mosquitoes. In fact, this knowledge prompted one scientist to speculate that because feet ranked first, maybe some of the compounds produced by bacteria on the feet were uniquely attractive, and because those same bacteria are responsible for the distinctive bouquet of Limburger cheese, mosquitoes might be attracted to the cheese—and they were!

Another experiment tested human twins and their attractiveness to mosquitoes. It turns out that identical twins are more alike in their appeal than fraternal twins. That could be attributed to differences in genetically determined skin chemicals, but in this era of the human microbiome, it could instead mean that identical twins harbor similar sets of microbes, which in turn collaborate to create a mosquito-attracting environment.

So all that is to say there is no answer to the question of why mosquitoes love some people more than others. To each her own, as they say. Or love bites. If you happen to be a mosquito-prone person—and even if you're not—you might want to know what really works when it comes to repellents. There's good news and bad. The good news: DEET works. It repels best and lasts longest. The bad news: there have been reports of adverse health effects. That said, many of these claims, especially in children, are unproven, and negative effects that are directly attributable to DEET are rare.

 **Did You Know . . .** In the 1980s a thirty-year-old man was admitted to hospital exhibiting psychotic behavior. For three weeks he had been applying DEET to his skin every day, then sitting in a sauna. It was suggested that the DEET caused the psychotic behavior, and anti-psychotic drugs cured the man, but it might have been simpler not to do that in the first place! It's not as if the sauna was teeming with mosquitoes.

Something like two hundred million doses of DEET are applied worldwide every year, and the number of true adverse reactions is a tiny percentage. Admittedly, it's easy to freak out when you see that your plastic sunglasses or watch crystal has melted because of contact with DEET, and it's even freakier when you know that DEET penetrates your skin and enters your body. Still, proven medical complications are extremely rare. If anything, there's a danger to overestimating the risks of DEET. In areas where disease-carrying mosquitoes are abundant, the risk of mosquito-borne disease is far greater than the risk of DEET side effects.

There's no question that DEET is effective, but how does it work? DEET has no effect on the mosquito's long-range detection of carbon dioxide, nor on its visual homing sense. But when a mosquito approaches you as its target, the DEET may interfere with its ability to detect lactic acid on your skin. As well, if DEET plugs into odor receptors on the mosquito's antennae, that could create confusion about where you are, making your bare legs, arms and other exposed body parts less detectable. At this point, the exact effect of DEET is still not known, but we do know for certain that it's the best repellent we have.

Did You Know . . . Years ago in Alaska, an experiment determined that DEET on the skin reduced the average number of bites per hour from a crazy 3,360 to . . . just 1! While DEET's repellent properties are well documented, other concoctions may work—to varying degrees. A comparative study published in the Journal of Insect Science tested DEET alongside a lemon-and-eucalyptus-oil formulation, several organic and "natural" repellents and Victoria's Secret Bombshell perfume. DEET came out on top, but the perfume was more effective than several of the other repellents tested.

A comparative study of mosquito repellents

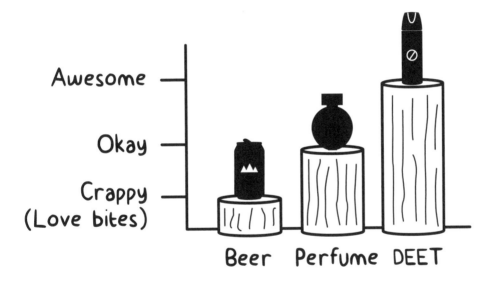

Why do lizards do push-ups?

You've probably seen this: normally lizards spend most of their time pretty much hugging the ground, but every once in a while they will launch into a frenzy of push-ups, or head bobbing, where they straighten their front legs and raise their chins. In some special circumstances they might actually engage in four-legged push-ups. Some species have a colored patch on their throat that makes the movement even more eye-catching.

Male lizards engage in these displays to signal their dominance over a territory. In many lizard species, male territories encompass several female territories, and the males desperately need to maintain their control over that space. It is a matter of life, death and reproduction. It's either do push-ups (which seem to keep other males at bay) or engage in hand-to-hand combat and risk serious injury. And

yet there is a cost to doing push-ups, just as there is if you're human: they take energy. So signaling just the right amount and no more is crucial.

Although it's difficult to prove, most researchers liken these displays to the vocal displays of birds and frogs, which are also signaling territorial rights.

Did You Know . . . For species like frogs, which croak their territorial rights, it's not simply an issue of who makes the most noise. Bigger frogs aren't necessarily the loudest. They may have a deeper call, though, with the pitch giving listeners clues to the size of the caller.

It's one thing to signal territory through push-ups, calls or other displays, but it's another thing altogether to ensure that the message is received. For birds, that sometimes means raising the volume of their calls if they're in a particularly noisy environment, which could be anything from the hum of city streets to the incessant pounding of a waterfall. Or they can alter the frequencies of their calls or songs to fit into a relatively quiet space in the audio spectrum around them.

For lizards doing push-ups, interference is not auditory but visual; something like vegetation waving in the wind might interrupt the message. In that circumstance, lizards will bow and stretch faster to separate their movements from the ones happening around them. And they may go even further than that. Experiments with lizard robots in Puerto Rico showed that when conditions for visual signaling were poor, as in low light or across large distances, lizards would issue an alert—a fast set of four-legged push-ups—before they actually got to the meat of the message. When the lizard robots were programmed to begin with the same alert, the real lizards paid attention sooner. In the absence of the alert, the signal that followed did not arouse the same attention.

Scientists suspect that much more information might be coded in these push-up displays than we are aware of, but so far, no one is certain of their full meaning. What looks to us like a tiny reptile doing a dance might convey all sorts of things to a fellow reptile. Who knows? Perhaps those messages are profound. Philosophical, even.

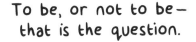

To be, or not to be—
that is the question.

Why do lizards shed their tails?

IF DOING PUSH-UPS TO ADVERTISE ONESELF didn't strike you as a little odd (even though you've undoubtedly seen such curious behavior at the gym), how about the idea of simply ridding yourself of a body part if you're attacked? Lizards are pretty fantastic at this. It's called autotomy—the defensive strategy of letting the tail go if it's seized by a predator.

Hmm...
I know
I'm forgetting
something...

When a snake attacks a lizard from behind, it bites at the tail right where it joins the lizard's body. But before the snake can swallow its prey whole, the rest of the lizard is gone, having fled while the snake was distracted by the wildly thrashing tail in its mouth. The lizard lives another day!

But as always, freedom has a price. Without a tail, a lizard's mobility is compromised, especially when climbing. Also, its mating display might be less effective. And in a temperate climate, losing a tail late in the season might mean a crippling loss of the animal's fat stores—stores that are essential to get the reptile through a winter of hibernation. The tail can be so crucial in this regard that some lizards take the chance of being attacked again and return to consume what's left of their own tail after a predator has abandoned it. Also, there's the problem of a second attack. It will be a sad tale next time when there's no tail for the lizard to drop.

The good news? Lizard tails regenerate. But that takes a lot of time and resources. So that's the balance that is struck: lose a tail to avoid almost certain death, only to raise the risk of dying months later.

 **Did You Know . . .** Some lizard species, such as the five-lined skink, have bright blue tails when they're young that later turn boring shades of brown. Why? It's possible that as the animal matures, camouflage becomes the preferred method of avoiding predators rather than tempting them with a bright blue morsel.

If a predator bit down on your arm and you ran away leaving your limb in its mouth, you can be sure you'd be in pretty dire shape. So what's different for a lizard? A closer look reveals how the unique anatomy of the lizard tail allows for autotomy. The tail is precut, divided into a set of fracture points. There really isn't a lot holding the tail together; even the muscles are fitted loosely, not firmly attached to each other. When the muscles around any of these preset fracture points contract, the vertebra breaks and the whole organ falls off cleanly, without leaving a bloody pulp behind.

Of course, escape is wonderful, but the really fantastic part of autotomy is what the tail does after it separates: the tip keeps flicking back and forth at pretty amazing speeds, and every once in a while, at least in some species, it suddenly lashes out. It is alive! Well, no. It isn't. But it does have self-contained neural circuitry and enough stored energy to keep moving for a minute or sometimes even more—long enough for the rest of the lizard, the living part, to run away.

It's curious that the frantic movements of the detached tail have different effects depending on the predator. Experiments with domestic cats showed that the vigorous movements of the detached tail confused or at least occupied the cat, allowing the tailless lizard to escape. With snakes, tail-lashing prompted concentration on subduing the tail's movements, thereby allowing the rest of the lizard to escape.

In the end it's not a bad bargain for predator or prey: the predator gets a nice fatty piece of lizard; the lizard gains freedom—at least for a while.

Why do some creatures throw their feces?

THERE'S A CATERPILLAR THAT DOES IT BRILLIANTLY, a penguin that is pretty good, and an ape that is truthfully more interested in the act of throwing than in what's being thrown. I leave it to you to decide which creature is more accomplished at shooting the shit.

Scatapult Scale

That caterpillars, at least some of them, can shoot their own shit like a projectile weapon likely comes as a surprise, but a closer look (not too close!) will convince you that (a) they do it in a sophisticated way and (b) it is essential for survival.

The skipper caterpillar is the best insect shooter. It's a leaf roller—that is, when it settles on a leaf to feed on, it pulls one side of the leaf over itself, just as you pull the covers over you when you get into bed. The only difference is that you don't anchor the covers in place with strands of silk as the caterpillar does. Once sequestered this way, the caterpillar is invisible, and predators that search for it using sight will not find it. But smell is another thing, and what gives the caterpillars away is the odor from their frass, which is a nice entomological word for excrement.

Some cool experiments in the lab have shown that wasps interested in a caterpillar meal will spend far more time inspecting rolled-up leaves containing frass than those without frass. So it pays the caterpillar to ensure that its frass is deposited far from its body. It's safety, not cleanliness, that has promoted this skill (and has earned this insect the nickname "scatterpillar.")

Researchers more than a hundred years ago noticed that skipper caterpillars launch their poop at a decent velocity several body lengths away, and in a variety of directions too, so they don't end up with a frass pile, making it obvious that there's a caterpillar in the vicinity. It took until the 1990s to clarify the actual mechanism by which they literally shoot the shit. The caterpillar's rear end is fitted with a complex of plates, "combs" and pressure vessels, all of which are synchronized to be able to send the shit flying.

Imagine, if you can, a fecal pellet ready for expulsion. It is maneuvered into position by a combination of fleshy collars on the creature's back end. These move the pellet into contact with the anal plate and the anal comb. The plate is the launchpad and the comb is the trigger, the crucial piece, hooked as it is on one of the collars until, with a rapid buildup of blood pressure (actually not blood but the insect version, hemolymph), the comb suddenly slips off its mooring and the plate lunges forward, sending the frass flying. It might sustain launch forces of more than one hundred Gs as it exits the caterpillar's body, at speeds in excess of six feet (two meters) per second. The mechanism has been charmingly described as similar to the game of tiddlywinks, where hard pressure on an object sends it flying.

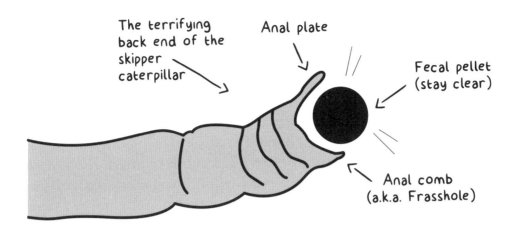

The terrifying back end of the skipper caterpillar

Anal plate

Fecal pellet (stay clear)

Anal comb (a.k.a. Frasshole)

It's a big jump, evolutionarily speaking, from a skipper caterpillar to the Adélie penguin, but that's pretty much how far you have to go to find another creature with at least a modicum of skill in this fecal volley game. Adélie penguins (and chinstrap penguins) eject their feces from the nest—a feat they accomplish simply by perching on the edge of the nest, rear end outward, bending forward, lifting the tail and letting it fly. But sadly, the forces and distances, at least with respect to body size, aren't nearly as impressive as the scatterpillar's.

Of course, not all poop is created equal, and in this case, the bird's waste is not solid but liquid, with the viscosity of something like olive oil. A typical nest-side shoot might be a little over a foot (about thirty centimeters), and unlike the compact little packages ejected by the caterpillars, this is more like a streak of sluice, colored pink if the last meal was krill or white if it was fish. It isn't as clear why penguins volley their feces, but the best explanation is that the birds spend a lot of time preening and cleaning, and getting rid of feces would be consistent with that. Incidentally, the streaks left behind radiate from the nest in all directions, but it's not clear whether the penguins do this on purpose.

And now, an evolutionary side step to one of the biggest land mammals: the hippopotamus. You don't have to venture far onto YouTube before you find videos documenting the delightful hippo habit of vigorously wagging the tail while defecating, ensuring that its feces fly off in all directions. I've seen this myself, and I can tell you it's impressive.

The mechanism isn't particularly sophisticated: if you were to fire blobs of mashed potatoes at a rapidly swinging pendulum (and I mean rapid), you'd get pretty much the same result. But why would you do that? And why does the hippo do it?

Most explanations are guesses. Spreading feces around might be a display of dominance by high-ranking animals; it isn't likely to be territorial, though, like dogs and hydrants, because male hippos' territories are in the water—a stretch of river, for instance. There'd be no point tail-swishing in the water because the feces would be taken away by the current anyway. Could it be just for amusement?

Chimpanzees love a good fecal throw—in the wild and in zoos. Zoo-goers know that there's always a good chance that chimps will throw their poop around, to the horror and/or delight of spectators. But they don't restrict themselves to poo, and their throwing skills say more about their brains than they do about what they're throwing. For ten years, researchers watched a chimp named Santino at the Furuvik Zoo in Sweden. Santino liked to express his general displeasure by throwing rocks, not poo, at visitors. But it was what he did when he wasn't throwing that intrigued the investigating scientists. When the zoo was closed, Santino would scout his territory for rocks that he could pile up in anticipation of new visitors arriving the next morning. He'd hoard these rocks, throw them when the visitors arrived, then stockpile at night once more—over and over in a cycle. Researchers drew the conclusion that his future-planning had a human-like quality.

This story intrigued me because I, too, was once the target of rocks thrown by an angry male chimp. I was in Japan, at the Primate Research Institute of Kyoto University, shooting for *Daily Planet*. The institute's director at the time was Tetsuro Matsuzawa, known worldwide for his work with both captive and wild chimpanzees. He showed me and my team a fantastic chimp named Ai that easily outscored me on a memory test. While Ai's memory scores were impressive, I also met an alpha male who had a pretty good underhanded throw.

As we approached, the alpha male retreated to the far side of his enclosure, looking anxious. Then he started shaking, picked up some stones, ran to his left and started hurling them at us. Matsuzawa thought it was hysterically funny. I was concentrating on not getting hit!

Whether it's rocks or feces, it's the throwing that interests scientists. In captivity, chimps seem to do it to vent, or for the entertainment value. It's the only time they can really exert an infl ence over visitors. In the wild, chimps throw at each other, but not usually with enough power and accuracy to injure. It's really about sending a message—not the most friendly one, either.

One recent study showed that the brains of chimps that throw well are visibly different, at least as seen through the eyes of an MRI. They have more elaborate sets of connections and faster, more efficient signaling on the side of the brain that controls the throwing arm, which is most often the left-brain-to-right-arm connection. And most interesting, the good throwers were also the better communicators, more adept at tasks that demand the chimp be aware of the researcher and communicate with that person.

In human beings, that cross-body control would also most often be a left-brain-to-right-arm connection. But we also have language, and there's an argument that the fine motor control of the lips and tongue for speech is a refinement of the neural circuits that control throwing. Maybe that's the reason the left side of the brain controls speech: it co-opted the already elaborate motor control systems on that side.

So throwing first, then speech (as opposed to language, which doesn't have to be spoken). Of course, it's also easy to imagine that gestures could provide the critical bridge between throwing and speaking. It's not the splatter of feces on your coat that is significant; it is the fact that the chimp is giving you a glimpse of how the human brain evolved. This is what sets the chimps apart from all the rest: yes, feces are thrown, but with intent.

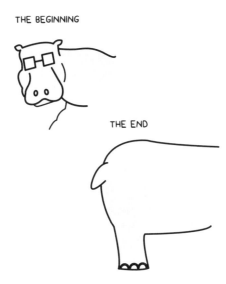

THE BEGINNING

THE END

Why do geese fly in V formation?

THE MOST POPULAR THEORY for why geese fly in a V formation has always been that in a such a formation, every goose gets a slight boost from the others—a literal updraft, a vortex of air that swirls off the tips of the wings of the birds next door.

If efficiency is the goal, why wouldn't geese instead fly lined up like a chorus line, each bird directly beside the other? The aerodynamics equations suggest that as few as nine birds flying in a chorus-line formation would gain 50 percent more range. But a V formation distributes the energy savings among all the birds a little more fairly: the goose in the lead benefits a little less (although still gets a bit of a kick), and birds farther back in the V fare better. That might be what makes it the go-to goose formation.

Some experts have suggested there could also be a social reason for the V: it's a perfect formation for keeping tabs of one's wingmen. But considering there are many ways for geese to maintain social cohesion, it still comes down to the V giving all the birds a lift.

A V formation is a robust thing, too. If a goose flies too far forward out of the V, she'll lose uplift and naturally fall back into line. If she drifts too far back, she'll get more lift and suddenly find herself surging forward into her proper place.

Did You Know . . . Fighter pilots experience a similar boost to geese by flying in a V formation. They save as much as 18 percent in fuel consumption by flying just behind the wingtips of the plane in front of them.

The positioning is crucial, because besides the much-coveted lifting vortex at the wingtips, there's also a "downwash" spilling off the back of the wing that would make it harder to fly. That's a good place to avoid. To work best, the V should have an angle of about 100 degrees—about the angle your little finger makes with your thumb if you stretch your hand as wide as possible.

For a long time, that's about all anyone knew: that the V was an energy-saver. But in the 1970s and 1980s a small team of dedicated naturalists decided to analyze the spacing of geese as well. They spent their days standing out in the cold, cold autumn, desperately hoping for a V of geese to pass directly overhead so they could photograph it and measure the angles and distances of the wings. Getting geese to fly where you want them to, when you want them to, is not exactly easy. This team got photographs all right, but interpreting the distances in the photos was challenging and contentious.

While they were figuring that out, new theories sprung up. One posited that the angle of the V was less important than the relative positions of the wingtips. Each bird's wingtips should be directly behind the wingtips of the bird in front. Researchers went back to their photos to check out this possibility and concluded that, yes, that's exactly how the birds were positioned. So maybe the secret was wingtip arrangement, not V angle. But there were still puzzling observations of geese flying in ragged Vs, or Vs with more birds on one side than the other. The question remained: Why?

The breakthrough came when two unrelated groups—European scientists interested in the aerodynamics of flight, and conservationists working with a rare bird called the northern bald ibis—found each other.

This ibis disappeared from the wild in Europe in the 1600s, and a group in Austria was trying to reintroduce them by hand-raising young birds and then guiding them along their historic migration route with an ultralight aircraft. This is almost exactly what Bill Lishman, also known as *Father Goose*, did in the 1980s and 1990s, when he trained geese, and later whooping cranes, to follow ultralights on a migration route. (His exploits were later dramatized in the movie *Fly Away Home*.)

Back in Europe, the new partnership between conservationists and scientists led to a great opportunity. Previously, the scientists had had a problem: the instrumentation they attached to the birds, which recorded position and wingbeat, worked only if the scientists could control where the birds landed. But how do you tell a flock of ibises where to land? Enter the conservationists. Becuase they led the flocks, they were able to guide the young ibises to land where the scientists wanted so that data could be downloaded at every landing spot.

And that data was gold. It showed that the ibises maintained exact spacing between each other in a V. But there was more than that: they also timed their wingbeats to take maximum advantage of the uplift provided by the bird in front. One of the researchers likened their efficiency to walking through the snow by placing your feet in the footprints of the person in front of you. There's a rhythm that maintains maximum lift, and these birds adopted it seamlessly.

But even more surprising was their adaptability when the formation shifted. These ibises never maintained a rigid V. Sometimes they even fell into single file. That should have been a bad thing, because they'd encounter the downwash off the wings of the bird in front, but they adjusted by automatically adopting a mirror-image wingbeat, in which up and down strokes were reversed. This limits the disadvantage of the downwash.

Interestingly, these birds were hand-raised by humans, so no adult ibises had taught them how to fly in a V. Still, they did it. It might be that they simply adjusted their positioning and wing strokes "on the fly" because it saved them energy.

So if flying in a V saves so much energy, why don't all birds do it? The aerodynamic theory suggests that the advantages accrue to large birds whose wings throw off big vortices. Sparrows, not so much. Smaller birds would have to fly extremely close to each other to get any uplift, and as far as anyone knows, they don't bother.

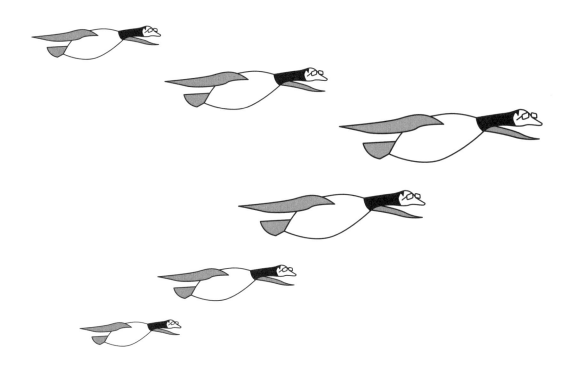

History Mystery

Why were Tyrannosaurus rex's arms so short?

Tyrannosaurus rex was not the biggest carnivorous dinosaur ever. As far as we know, that distinction goes to the peculiar swamp-dwelling *Spinosaurus*. *T. rex* isn't even the biggest and fiercest of our imaginations—that's *Jurassic World*'s CGI colossus, *Indominus rex*. But who cares? It's still *T. rex*, and whether it occupies the all-time heavyweight throne or not, it was the ruler of the Cretaceous period. Besides, I challenge you to name another dinosaur that inspired the name of a rock band.

The all-time heavyweight champ is... Spinosaurus!

A typical *T. rex* could reach 40 feet (12 meters) in length, measure 13 feet (4 meters) at the hips and weigh nearly 7 1/2 tons (7 tonnes). To grow to those dimensions, teenage *T. rex* would go on a weight-gaining spree, putting on 1300 pounds (600 kilograms) a year from ages fourteen to eighteen. It'd go on to live into its mid-twenties. Even at that size, it could manage speeds of roughly 18 miles an hour (8 meters a second), making it a fierce predator. Its astounding bite force was ten times as powerful as that of an alligator, making it the most powerful ever seen on Earth. But there is one feature of the *T. rex* that has always puzzled experts: its tiny forelimbs.

On a full-sized *T. rex*, the forearms were only about three feet (one meter) long. There are basketball players with a reach like that. When he described the first skeleton of *T. rex* in 1905, the American paleontologist Henry Fairfield Osborn—having recovered only the dinosaur's upper arm bone, the humerus—imaginatively reconstructed the beast with a long, three-fingered forearm. The first complete set of *T. rex* forelimb bones wasn't discovered until 1989, and those bones confirmed what many had suspected from studying related species: *T. rex* had absurdly short arms—too short even to reach its mouth.

The immediate question, then, was what were those tiny arms good for? Some dismissed them as vestigial, envisioning a progressive shrinking from the full-length prey-catching arms of smaller carnivorous ancestors to the stunted forelimbs of the *T. rex*. The problem there is that however short they might have been, the *T. rex*'s forearms were heavily muscled. Inferring power from fossil traces is one thing, but it's also possible to create mathematical models of the *T. rex* forearms, plug in known values and calculate what they were capable of. Both approaches arrive at the same result: the forearms were powerful and fast.

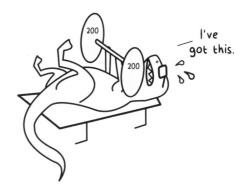

In fact, to say they were heavily muscled would be an understatement. Judging from the thickness of the arm bones—three times the size of an average person's—and the huge areas on them for the attachment of muscles and tendons, we can conclude that the arm was at least the thickness of a human thigh. The biceps attached farther down on the forearm than ours, giving the *T. rex*'s arm more leverage, and they had help from the surrounding muscles, especially the huge shoulders. With all that power behind them, each arm alone could curl close to 450 pounds (200 kilograms)! If evolution was gradually marginalizing the arms, though, it wouldn't make sense to retain that kind of musculature. The arms must have had a role to play.

 TRY THIS: Straighten your arm in front of you, then touch your shoulder. That one movement takes your elbow through 165 degrees. Because of all the muscle on its forearm, the *T. rex* could manage only 45 degrees, with very little side-to-side movement. It just wasn't built to do anything more than simple flexing.

Osborn speculated that the muscle on the arms might have been involved in sex—but don't forget, he didn't know how incredibly short those arms were. So what exactly did these powerful but apparently undersized pieces of machinery do? Some scientists used the shortness of the arms as evidence that the *T. rex* didn't run very fast. If a *T. rex* had tripped, its little arms wouldn't have been able to break its descent, and such a fall for a beast that size would likely have been fatal: given its weight, it would have slammed into the ground with six times the force of gravity and broken bones in the process. But this theory was questioned by referencing other animals. Giraffes can sprint 30 miles per hour (50 kilometers an hour) and would likely break a leg if they fell. Ostriches move very fast, too. Neither animal has arms to break a fall, and so there's really no reason to think *T. rex* was any different.

My favorite explanation was offered by paleontologist Barney Newman in 1970. Newman envisioned a *T. rex* "in a position of rest," lying face down with its jaw on the ground. How would it get up? The power would have to come from the giant hind legs, but if they pushed with no counterforce, the animal would skid forward on its belly. Instead, Newman suggested, the first thing the dinosaur would do is dig its front claws into the ground and do a push-up. That would allow it to rock backward and upward, eventually standing on its hind legs.

Newman's idea hasn't exactly been shot down, but it hasn't prevailed, either. Much more emphasis these days is placed on the notion that the *T. rex*'s mini-arms were somehow useful in feeding. There is evidence that they were used in hazardous situations, because many of the surviving fossil *T. rex* arm bones have been chipped, cracked and broken, apparently by excessive forces. If all those arms were used for was propping the animal up, it's hard to see how they would incur such damage. At the same time, it's also challenging to imagine a *T. rex* using those tiny, bulging arms to help position and subdue a gargantuan *Triceratops* so that its giant jaws could make the kill.

There's always been some controversy about whether the *T. rex* was a mighty hunter or simply an oversized scavenger, the multi-ton buzzard of the dinosaur age. Consensus seems to be building in support of it being a hunter.

In fact, according to Canadian paleontologist Phil Currie, *T. rex* was a pack hunter, like the wolf. The best evidence of this is the fact that *T. rex* had a brain three times the size of other dinosaurs—pack hunting requires smarts to co-ordinate the activities of several animals simultaneously. A set of tracks found in British Columbia of several tyrannosaurs (either *T. rex* itself or that or a relative) walking together further supports the theory.

If *T. rex* was smart, though, there were limits to its intelligence. There is one famous fossil of a *Triceratops*, the three-horned dinosaur with the huge frilled shield on its head, where one horn has been broken off and part of the frill has been chewed. *T. rex* is the only possible candidate for the attack. Is it intelligent to attack such an animal—which would have been the size of an elephant—from the front, where all of its weapons and armor were? Well, no, it isn't.

You'd think that with two decades of the *Jurassic Park* brand under our belts, we'd have a fairly accurate depiction of *T. rex* in our minds. Especially in kids, who devour anything dinosaurian. Apparently not. A trio of educators at Cornell University tested both college and school-age students for their understanding of *T. rex*'s posture. When *T. rex* was first unveiled to the public, in 1905, it was depicted standing upright, dragging its immense tail along the ground (see Exhibit A).

EXHIBIT A

That was no more than a guess—and it was wrong. When scientists started to dig deeper, they realized that tail-dragging made no sense, not to mention that none of the fossil trackways of carnivorous dinosaurs like *T. rex* showed any signs of a dragging tail. In fact, an upright posture would likely have damaged several of *T. rex*'s joints, including the hips. Instead, think of *T. rex* as a horizontal animal, its head extending out front and its tail straight out behind. The animal likely swaggered a little as it walked, swinging its tail from side to side to balance the huge weight shifts caused as each giant hind leg stepped forward (see Exhibit B).

EXHIBIT B

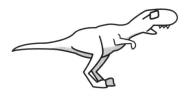

To its credit, *Jurassic Park* portrayed the dinosaur the right way (although *T. rex* existed in the Cretaceous period, not the Jurassic), and that was in 1993. A full generation since then has had a chance to adopt the correct posture of *T. rex* for its shared depictions of the animal. But the Cornell experiments suggested the exact opposite had happened. In that study, 111 students were asked to sketch a standing *T. rex*; the average angle of the spine relative to the ground depicted was more than forty-five degrees, closer to the 1905 original and far from today's interpretation, which would be closer to zero degrees!

It seems unbelievable that a dramatic change in the posture of the world's most famous dinosaur has failed to penetrate the minds of its biggest fans. But there it is. I blame Barney.

Part 3
Supernatural

What is déjà vu?

CHANCES ARE YOU'VE EXPERIENCED DÉJÀ VU BEFORE, in which case, one thing is for sure: you remember what it was like. Ironic, because the essence of déjà vu is a failure to remember. Déjà vu is that sudden vivid feeling that wherever you are and whatever you're doing, you've been there and experienced that situation before, even though you have no actual memory of it.

Technically, there are several slightly different versions of déjà vu, including déjà vécu (the feeling of having lived through an event), déjà senti (the feeling of having already felt a particular emotion), déjà visité (the feeling of knowing a place you've never visited before) and déjà pensée (the feeling of having had the same thought before). (Do writers have déjà écrivée?) Despite the shades of gray, each variation is really about the same sort of experience.

Lest we forget!

Consider this example of déjà vu. You walk into a room full of people, and as your attention is pulled one way, then another, you briefly notice a lamp on a table.

Later, as the crowd thins, you get a better look at the lamp and think, "I've seen that before," which expands to overwhelm you with the feeling that you've actually been in that room before. It might be only a momentary occurrence—as that instant feeling of familiarity fades quickly—often because you're busy trying to remember why it's happening in the first place. You think, "Where was I when something like this happened previously? What is it that's so familiar about this situation?"—Rationally, you know that you haven't really been in that situation before, but something's going on in your brain to make you think you have.

Déjà vu is a frustrating psychological occurrence. It is estimated that roughly two out of every three people have experienced it. Despite how common it is, déjà vu is poorly understood. Even the experts—psychologists, neuroscientists, psychiatrists—acknowledge that they don't have a precise handle on it, but they would also say that they're getting closer, dispensing with old ideas and refining new ones.

Given the lack of an exact scientific explanation, it isn't surprising that we have poets and authors to thank for many of the most vivid descriptions of the déjà vu experience, as well as some of its worst explanations. In the "best description" category, here's David Copperfield, in Charles Dickens's classic tale: "We have all some experience of a feeling, that comes over us occasionally, of what we are saying and doing having been said and done before, in a remote time—of our having been surrounded, dim ages ago, by the same faces, objects, and circumstances—of our knowing perfectly what will be said next, as if we suddenly remembered it!"

If you've experienced déjà vu, you'll realize that Dickens's observations are right, but it's the part about knowing what will come next that's really intriguing. How many people actually feel that level of clairvoyance? And how long does that feeling last? If someone says something and you think you know what he or she is going to utter next, then you should be able to predict the next statement. But there is no proof that the feeling of a prediction leads to actual predictions.

Dickens's reference to "dim ages ago" may be, as countless others have suggested when it comes to déjà vu, evidence of reincarnation. Of course, reincarnation would be a simple, straightforward explanation . . . if it weren't for two major stumbling blocks. First, there's no credible proof that reincarnation is possible. And second, if one did previously experience an identical situation "dim ages ago," how likely is it that it would be remembered?

Bean there, done that...

Think of a conversation you could have had in your living room today. Then imagine a previous life in which you might have been sitting in front of a campfire, dressed in your filthy, lice-ridden woolens, or if you were of higher social standing, sitting in front of the fireplace dressed in your fashionable, lice-ridden woolens. It's unlikely that the conversation you had in the dim past would have been about binge-watching *Breaking Bad*. Rather than reincarnation being the explanation for déjà vu, perhaps déjà vu led to the very idea of reincarnation in the first place.

Other paranormal explanations run through déjà vu. In the words of "thought energy expert" Jeffry R. Palmer: "Far from discounting the study of the paranormal, the recent theories describing déjà vu experiences as electro-chemical misfiring in the brain . . . highlight the importance of continued research into paranormal phenomena."

Actually, "electro-chemical misfiring" isn't a justification for researching the paranormal; it

simply means that there's something odd going on in the brain. The question remains: What is that something?

Nearly a hundred years ago, psychologist Edward Titchener suggested that déjà vu was a result of a broken sequence of thoughts. Imagine, he suggested, you look both ways before crossing the street and just for a second your attention is caught by the display in a store window across the road. You then cross the street, but when you catch a second glimpse of the store window, you think, "I've been on this street before." Titchener's argument was that the original left-right look is separated in your mind from the actual crossing by the images of the store window, which makes you think that the first experience (the looking) is a much earlier and fuller experience in the past than it actually was.

Another possibility is that the processes of memory making and memory retrieval, which most of the time are independent, occasionally happen simultaneously, resulting in a feeling of remembering a situation you're actually experiencing for the first time. This was once likened to having the record and playback heads of a tape recorder active at the same time. But the tape recorder analogy isn't a great one anymore (fewer and fewer people even know what one is!), and the theory isn't that popular either.

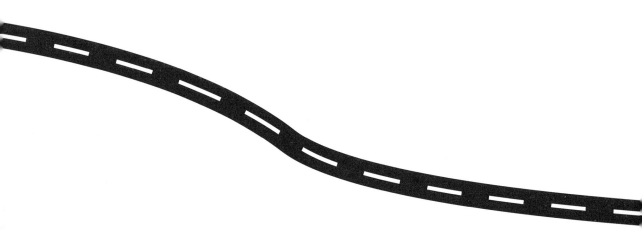

There's another explanation that is the opposite: this one relies on two normally synchronous processes separating momentarily. One example—and this is a hypothesis only—is that a sense of familiarity and the retrieval of the memory associated with that familiar situation usually happen at the same time. But if the timing is disturbed and the feeling of familiarity jumps ahead of the memory, you're left thinking, "I know this place," with no relevant memory to support the feeling. Result: déjà vu.

Finally, another explanation hinges on features of a place that trigger the déjà vu experience. Imagine you walk into a room in which you find an old television exactly like the one your grandfather owned. The TV is familiar, but it's out of context, so you can't put your finger exactly on where you've seen it before. Instead, you think you've been in the room before.

TRY THIS: I'm sure that you have driven some distance on a highway either without remembering it because you were listening to the radio or, conversely, without hearing what's on the radio because you were concentrating on driving. That latter situation is a perfect opportunity for déjà vu. You've heard what was said, but you didn't notice it or weren't conscious of it. If you were to hear one of them again, it's possible that it could be dredged up from wherever it landed in your brain. It's quite likely it's there somewhere, and there's a lot of evidence that these unconscious thoughts influence us all the time. Try to remember what songs you heard on the radio in your last car trip. Can you remember them all?

Although specific objects can trigger déjà vu, the experience is not necessarily tied to or triggered by objects. Anne Cleary at Colorado State University has done many experiments showing that déjà vu can be triggered in the lab. For one, she created two sets of pictures that had similar layouts but very different subjects and details. She started by showing student volunteers one set of images. When she showed them the second round of images with the structurally similar scenes the students were convinced they'd seen the pictured place before. It turned out that it was the configuration of the items in the picture, not the specifics of those items, that stood out in the students' minds and led to the déjà vu. Dr. Cleary has extended these experiments by using virtual reality technology, and when students are immersed in the scenes presented to them, the déjà vu feeling can still be created. The experience is called feature-matching, and Cleary says it can "produce familiarity and déjà vu when recall fails."

Castle

Parliament

Burger Manor

Of course these experiments are at best only a partial explanation of déjà vu. The composition of a scene is one thing, but a real-life, complete experience includes people, conversations and many other sensory details, none of which are included in Cleary's experiments. Psychologist Bennett Schwartz described experiencing Cleary's feature-matching. He had a powerful déjà vu while touring a castle in Scotland. When he was in the gift shop after the tour, he saw photos from a movie shot in the castle. He then recalled that he had watched that film five years before. Presumably, it was the memory of the movie scenes that triggered his déjà vu.

Science _Fact!_ *For some reason, castles are excellent places for déjà vu. Even before Schwartz, at least five déjà vu experiences documented in the literature happened in castles—English, Scottish and German—from the 1860s to the 1950s. Do we all have vivid mental images of castles even if we've never been to one— images that could set off a déjà vu?*

So far, there's no hard data to support any one explanation of déjà vu. Something is going on in the brain, and although the specific nature of that glitch hasn't been proven, some neurological conditions are associated with a much greater frequency of déjà vu. One in particular is temporal lobe epilepsy. Many sufferers experience déjà vu shortly before they have a seizure. Temporal lobe seizures are usually the result of a small area of scarring in the brain beside and behind the ear. It has been suggested that small disruptions in that region might produce the déjà vu effect, even in people who don't have epilepsy.

Some researchers have tried to determine who among the general population is prone to déjà vu. It's extremely rare in children below the age of eight or nine, but its incidence rises rapidly through childhood before starting to decline later in adulthood. And it seems to happen more often when people are stressed, tired or anxious. In 2014, researchers reported a remarkable case of a twenty-three-year-old man who experienced déjà vu as a result of chronic anxiety. His case was so severe that he stopped watching TV and reading newspapers because he felt he had seen and read everything already.

A handful of other people have reported such overpowering déjà vu. One woman stopped playing tennis because she felt she could anticipate the result of every rally. The apparent champion of déjà vu frequency is an otherwise mentally healthy individual called Mr. Leeds. In the 1940s he tracked his déjà vus for a year and reported 144 of them—all in incredible detail! Still, many of these are extreme cases accompanied by other psychological disturbances.

Did You Know . . . Educated people who travel are more prone to déjà vu. It's possible this is because they've stored more images and memories that resemble new encounters. Also, dreamers, especially those who remember their dreams, are more likely to have déjà vu.

We can't leave this subject without acknowledging déjà vu's evil twin: jamais vu, the feeling that you've never been in a place that you're very familiar with. It's been suggested that jamais vu is like "word blindness," where if you read or speak a word over and over, it starts to become meaningless, to lose all familiarity. Take the famous *Friends* episode called "The One with the Stoned Guy." Jon Lovitz, who is playing a restaurateur in search of a new chef, arrives at Monica's apartment to sample her cooking. It's clear to the others in the room that Lovitz's character is high, and when Monica mentions the tartlets, Lovitz excitedly repeats the word three times, as though he has something else to say on the matter, before announcing, "The word has lost all meaning." If you take that sort of mindless repetition that's devoid of meaning and multiply it by ten, you get closer to the experience of word blindness, or roughly, jamais vu.

Do we dream in color?

BEFORE YOU IMMEDIATELY REJECT THAT QUESTION AS RIDICULOUS, think for a moment. If you can remember what you dreamed about last night, go over the events, objects and people that made up the dream and try to recall what color they were. It's much harder than you think. Sometimes it's apparent only when a color plays a key role, such as in a dream about sitting in your car at a stoplight and accelerating only when the light turns green. And even then, was that light really green? Or did you assume it was green because it was the bottom of the three lights?

If we don't dream in color, what about black and white? That question is just as difficult to answer. It does seem unlikely, given that our waking lives are dominated by color, that we'd revert to black-and-white dreams. Even more interesting: What if dreams are neither color nor black and white, but just . . . nothing? Think about the last novel you read: Did you color in every person and scene in your head, or did you simply process them as people or things, without visually imagining them in detail?

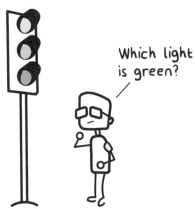

Which light is green?

125

Why ask these questions? In 1940s and 1950s surveys, when people were asked, "Do you dream in black and white or color?" a majority said black-and-white. In one 1942 survey of college students, 71 percent of respondents reported never dreaming in color or only rarely. No more than 10 percent claimed their dreams were in color.

 TRY THIS: When I put out the question "Do you dream in color?" on Facebook, the responses were mixed: most people said they dreamed in color, and a few claimed those colors were incredibly vivid. Some thought you choose to color something only if it was important in the dream, or even that colors became an issue only when you were remembering or recounting the dream. Also, three people asked themselves in the middle of a dream that night whether that dream was in color. In all three, it was! What about you?

But then things changed in the late 1950s and early 1960s. According to published reports, anywhere from 50 percent to 100 percent of those surveyed reported dreaming in color. As for present-day surveys, 65 percent of respondents claim to dream in color and 25 percent dream in both black and white and color. No more than 5 percent report dreaming in black and white only. And the other 5 percent? They don't know. That is a dramatic change over sixty years—so what happened?

There are two possible answers. One is that sixty years ago we actually did dream in black and white, and now we dream in color. But it just doesn't seem possible this could be true. Fundamental brain activities like the perception of color and dreaming are extremely unlikely to change in just a few decades.

If dreams have always been in color, then the only other explanation is that people's testimony changed. People were having color dreams but claiming they were black and white. Seems odd, but there is some evidence this really happened.

Philosopher Eric Schwitzgebel has examined dream reporting over time and shows that dreaming in color has been around for millennia. Aristotle mentions dreaming in color, and half of the dreams that Freud analyzed in detail in *The Interpretation of Dreams* mentioned color. Schwitzgebel also makes the connection that before the late nineteenth century, colored dreams were compared to paintings or tapestries.

And there's the hint: maybe it's not so much what actually happened in a dream, but what the person reported as happening. And that's different. If visual surroundings, like tapestries and paintings added color to dreams in the past, what replaced them in the 1940s? Black-and-white movies, black-and-white photography and newspapers were the dominant media in the very years when people reported they dreamed in black-and-white. Could those media have directly influenced the recollection of dreams? You can certainly be skeptical (after all, people were still living in a colored world). But you can't ignore the fact that as dream reports change with time, so do the media.

The black-and-white era began to disappear in the late 1930s, when movies like *The Wizard of Oz* and *Gone with the Wind* created a Technicolor sensation. But only in the 1950s and 1960s did color TV and color photography become common in middle-class homes. Today, even newspapers are in color—is this proliferation of color media why we now claim we dream in color?

It is scarcely believable that we could be swayed to adopting color dreams. But a 2003 study of Chinese students lends support to the claim that the media influence our dreams. The students varied dramatically in their familiarity with technology. Sure enough, those who had had the briefest exposure to color media, such as color TV, reported the fewest colored dreams.

Even with evidence like that, the whole idea seems strange. Try it yourself: I bet you'll see that it's hard to remember the colors of dream furniture, cars or clothing; you might not remember any color. Maybe that's not surprising: it's possible color materializes only when attention is cast on something. If that's true, dreams could be both color and black and white. And if we really do dream in color, then those who opted for black and white back in the 1940s and 1950s had some powerful influence acting on them. If it wasn't media, what was it?

So what's in store for us? Black and white versus color will look like Beta versus VHS in the fully experiential future: VR gone wild, with all five senses engaged in whatever production you're immersed in. Imagine what experiences like that might contribute to dream imagery.

Can we really tell when someone is staring at us?

Have you had this experience? You suddenly feel that someone is staring at you from behind. You turn to check and they are in fact looking right at you. How could you have known they were doing that?

For centuries, people have believed they can feel stares before they see them, but it wasn't until about a hundred years ago that anyone attempted to explain this through the lens of science rather than magic or parapsychology. In 1898, British psychologist Edward Titchener pre-

sented an explanation that is still favored today. He began by observing that people are concerned about being viewed from behind, and will, if seated at the front of a crowded auditorium, constantly adjust their hair, or brush their collar, or even glance behind them. In turn, those

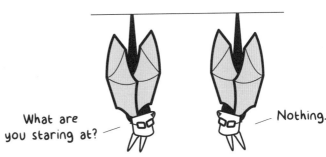

What are you staring at?

Nothing.

movements attract the attention of people behind them. So when the person in front happens to look back, she will inevitably see someone looking at her, and will be convinced she sensed that stare. Titchener's theory cleverly turned the situation on its head (literally). And the timing works: While it takes about a second to turn your head, it takes the starer only about a fifth of a second to shift his or her gaze to you.

In 1913, John Edgar Coover followed up on Titchener's work. He had a "starer" stare at a person based on a roll of a die. When the starer rolled odd, he was permitted to gawk at a "staree." When he rolled even, no gawking. When he stared, he was to stare for fifteen seconds straight. Meanwhile, the staree was tasked with guessing, without turning around, when he was being stared at and when not. Coover's results showed that the person being stared at could not accurately guess when it was happening. The guesses—at chance levels, 50.2 percent—were no better than results obtained by simply flipping a coin.

In this study and others, some issues continue to challenge clear answers to the question. Did the starer randomize the timing of the stares? If not, then positive results could be the result of the staree starting to anticipate when he was being stared at by sensing a rhythm. The staring had to happen at random intervals for the test to yield accurate results. Nor could there be any sounds involved—no shifting in the chair, no rustling of a shirt, no clues whatsoever of the starer's gaze.

In 1993 at the Institute of Transpersonal Psychology in California, William Braud conducted some eye-opening research. Braud felt that the problem might be asking people to say whether they were being stared at. Their intense concentration on stares, together with uncertainty about exactly what being stared at feels like, might blind them to subtle, instinctive signals. So Braud equipped his volunteer starees with electrodes that would record physiological arousal. This apparatus was something like the polygraph.

He put the starer and the staree in different rooms, allowing the staree to simply sit quietly for twenty minutes in front of a camera without being required to guess if he was being stared at or not. The starer, meanwhile, was looking at a TV monitor aimed at the back of the target's head, and he or she either stared or not, depending on cues given. The routine was ten stares, ten non-stares, each lasting thirty seconds, randomly ordered.

The results were sensational. There was a significant timing correlation between strong physiological responses in the staree and the actual stares from the person in the other room. Not only that: there was strange evidence of the two people being connected physiologically. As the starer became more comfortable with the odd situation of staring at the backs of people's heads, the starees' arousal levels declined. All of this, remember, with two people in separate rooms with a camera and a monitor. It boggles the mind.

As convincing as it seemed, this experiment is not the final word on the existence of a stare-guessing instinct. A similar set of experiments was conducted in the mid-1990s by the two-person team of Marilyn Schlitz, a well-known believer in psychic phenomena, and Richard Wiseman, a well-known non-believer. To their credit, they decided to partner up for the experiments. Using a setup similar to Braud's, they performed tests in Wiseman's lab in England and in Schlitz's lab in California. All were done the same way, but astonishingly, when Schlitz was the starer, the staree (whoever it was) predicted stares at a rate far exceeding the rate of chance. However, when Wiseman was the starer, the staree predicted at rates no better than chance. Both Schlitz and Wiseman were confounded by these results. So they took the next step: they designed an experiment in which they would switch roles. One would greet the staree at the beginning of the experiment, while the other would do the staring; then they'd switch. The justification for this was the admittedly slim possibility that Marilyn Schlitz's high results were somehow due to her establishing better rapport when she met the starees. In this new round of tests, she would sometimes meet the staree and sometimes not.

The experiment failed to find any evidence of psychic staring, no matter who stared or who greeted. This left the team with the difficult situation of having conducted almost identical experiments three times, twice with at least partial positive results (Marilyn's), and now once with negative results for both Schlitz and Wiseman.

Allow me now to back up a bit, to the 1980s and renegade British scientist Rupert Sheldrake, who stared deeply into the heart of the problem. He is best described as anti-establishment, arguing for phenomena that, according to orthodox science, couldn't happen. Knowing you're being stared at was a phenomenon he set out to prove was real, and he even speculated on how it would work.

Sheldrake argued that the explanation lies in the mechanics of vision. Science sees vision as a sequence: light enters our eyes, triggering neural signals in the brain, and the patterns of light are analyzed to form images. Light enters the eye—nothing leaves it. That's what every single piece of evidence about seeing supports. But Sheldrake argued that this is only half the picture, that our eyes also send out unperceived signals, or "fields," and these affect what the eyes are looking at. There is only one problem with Sheldrake's theory: there isn't a shred of evidence to support it.

A scientist, outstanding in his field.

The word "field" can be taken to mean just about anything, and because science is full of fields—gravitational, electromagnetic—the word has the ring of truth about it. Until there's real, solid evidence of "psychic" staring, in a controlled environment yielding consistent results, the idea of stare prediction remains an intriguing idea only. The eyes may have it, but so far, we have no proof.

Does subliminal advertising work?

THE QUICK ANSWER IS THIS: under very tightly controlled conditions, subliminal advertising can work. The catch is that the circumstances have to be so tightly controlled that it likely isn't worth taking the time to create them. Without the ideal conditions in place, subliminal advertising just isn't all that effective.

What's on your mind?

Oh, / nothing...

NOTHING

A subliminal message is one that is presented on a screen so quickly that you're not consciously aware of it. These messages can be measured in thousandths of a second. Although these sorts of messages are too short to be consciously recognized, they can still have an effect on your brain, because most of what's going on in there is unconscious. Your consciousness—your awareness of what's going on in your own head—is actually only a tiny fraction of the second-by-second activity in your brain.

Both lab experiments and analyses of consumer behavior suggest that more often than not, we're acting on behalf of our unconscious mind. In one experiment, a group of students was exposed to a long list of words, some of which suggested old age, such as Florida, bingo and forgetful. On a similar list for a second group, the "old age" words were swapped out for ones that don't suggest age, such as California, awkward and chess. The study found that the students who had been given the "old age" terms walked more slowly when they left the room, but they denied being consciously aware of their slow gait. To the researchers, that seemed to prove that the students' unconscious minds had picked up on the cues in the terms and were guiding their actions accordingly. But a swirl of controversy—still unresolved—broke out when subsequent researchers said they'd been unable to reproduce those results.

A less contested experiment in the Netherlands showed that purchases of major items—things like cars—don't necessarily require a lot of thinking or analysis. In the test, researchers gave volunteers a set of characteristics of different cars, and then they either distracted the buyers with mental exercises or allowed them to quietly contemplate the purchase. Those who had been distracted—meaning they weren't able to think about the purchase—made the most log-ical choices, while the people who thought about the purchase long and hard selected lower-quality vehicles. The study showed that when it comes to buying items with fewer features to consider, such as oven mitts or toothbrushes, it's fine to use your conscious mind to decide what choice is best. But when the decision is a complicated one, the unconscious mind might actually be more reliable.

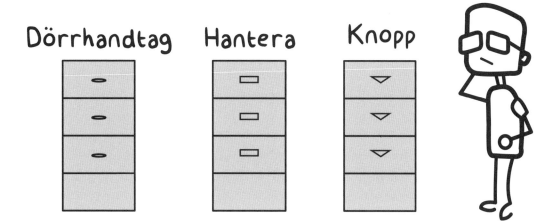

Dörrhandtag Hantera Knopp

Subliminal advertising messages seem like a logical step from there: persuade the unconscious mind to buy the advertised article. The research shows such messages can work, but only under tight constraints. In one experiment, researchers showed a group of volunteers subliminal ad that displayed the brand name of a soft drink. The scientists hoped the ad might encourage viewers to choose the advertised product over a competitor's. What they found, though, was that the messages worked only if the volunteers were thirsty. Not only that, but if the advertised brand was already the person's preferred drink, the subliminal message didn't increase his or her desire for it. The experiment did find that if people were thirsty and they didn't have a strong preference for any particular drink to begin with, they would tend to choose the drink that was subliminally advertised. But the conditions in the lab were carefully controlled—nothing like the real experience of either TV watching or moviegoing—and the researchers were forced to accept that the motivation to buy or choose a product has to exist before a subliminal ad has any effect.

There have been many attempts to harness the supposed power of subliminal advertising. The first was in 1957, when an ad man named James Vicary held a press conference at which he announced that his company had invented a new way of influencing consumers. He reported that he'd run a six-week experiment at the Lee Theater in Fort Lee, New Jersey, in which, throughout the hit movie *Picnic*, two alternating messages—"Drink Coca-Cola" and "Eat Popcorn"—had appeared for 1/3000 of a second, every five seconds. Vicary claimed that over the course of the test, forty-five thousand people were exposed to the messages, and he boasted that the marketing increased sales of Coca-Cola at the theater by 18.1 percent and sales of popcorn by an incredible 57.7 percent.

His announcement was a bombshell at a time when ideas like brainwashing and hypnotism were on everyone's mind. Unfortunately, the air was completely let out of the story when, in 1962, Vicary admitted that he had made up the entire study in an effort to drum up business for his marketing agency.

But that didn't stop others from claiming that they had developed their own methods of subliminally influencing people, notably the late Wilson Bryan Key. Key published the book *Subliminal Seduction* in 1973. The following year's paperback edition featured a great cover: a photo of a tumbler filled with a mixed drink, ice cubes and a twist of lemon. It was a commonplace image, but the caption for the photo read, "Are you being sexually aroused by this picture?" Key claimed that thousands were. The book was a bestseller.

Did You Know . . . In 1978, the heavy metal band Judas Priest was accused of including a subliminal message in one of their songs. Groups of concerned citizens complained that the band's song "Better by You, Better Than Me" included a hidden lyrical message saying, "Do it." The band ignored the complaints, but then, in 1985, two young men in Reno, Nevada, attempted suicide. Their parents claimed the song made them do it, and they took the band to court. But when evidence was introduced that both boys had significant emotional problems and had talked about committing suicide, the case was dismissed. Wilson Bryan Key testified, but he did not verify whether there was any meaningful subliminal advertising in the song. He did, however, tell the court that subliminal messages could be found in Ritz crackers (he claimed the holes were arranged to spell the word "sex"), in the Sears catalogue and on NBC news.

Does Bigfoot exist?

MOST OF THE SUBJECTS IN THIS BOOK qualify as science, but sometimes I like to write about "science." "Science" includes topics that have a scientific veneer, but a little scraping and sanding reveals there's nothing much underneath. UFOs, spontaneous human combustion and the Loch Ness monster are all perfect examples. Some of these "science" topics have quietly vacated the headlines, but others linger, no matter how little evidence supports their existence. I am fascinated—not because there's evidence validating these topics but becuase of people who labor long and hard to maintain their belief in them despite that lack of real evidence. One of my favorite "science" topics is that great man-ape of the Pacific Northwest, Bigfoot, or the name I prefer, Sasquatch.

It would require several books to give proper due to the Sasquatch legend, but I'll do my best here to probe into the "evidence" of its existence, knowing as I do that many believers will think I've shortchanged the animal terribly.

After all these years, the prime piece of evidence remains the Patterson-Gimlin film of 1967. That year, Bob Gimlin and Roger Patterson were drawn to Bluff Creek, California, by rumors of mysterious large footprints found in the forests. The pair took their camera with them in the hopes of spotting the creature that had made the prints. And lo and behold, they did!

The film they shot depicts a large human-like, two-legged furry animal walking unhurriedly from left to right across a clearing in a forest. The footage is pretty shaky. Patterson claims his horse reared at the appearance or scent of the creature and threw him. Only when he regained his equilibrium was he able to aim the camera properly and get a better shot of the beast.

A skeptic's first thought at seeing the film is that this was a guy in a gorilla suit, albeit a pretty convincing one. Fortunately, I was lucky enough to attend a two-day Sasquatch conference at the University of British Columbia Museum of Anthropology in 1978, where the gorilla-suit claim and other Sasquatch-related clues were examined closely. At the conference, Russian and North American Sasquatch acolytes dissected, slowed, paused, 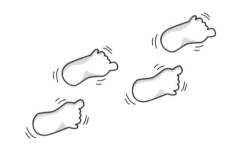 rewound, reshowed, worshipped and critiqued the Patterson-Gimlin film, all in an effort to come up with definitive evidence that this was some new creature, not a human in disguise.

Some said the figure was too broad-chested to be a human; subsequent research has shown that not to be true. One expert claimed that if the film had been shot at twenty-four feet per second, the thing was walking like a human, but if it had been shot at sixteen feet per second, it wasn't. Patterson wasn't sure whether the camera had been set at sixteen or twenty-four feet per second. How handy for the perpetuation of the story that there was this uncertainty!

Next on the docket: an examination of the movements of muscles underneath the skin of the creature. It was argued that rippling muscles would be impossible for a man wearing a loose-fitting gorilla suit. The rotation of the torso as the animal glances back at the camera couldn't be accomplished believably with a suit, and the arms were too long to be a human's.

Much discussion was given to the fact that the creature had "pendulous breasts" and therefore must be a female.

John Napier, an orthopedic surgeon, paleoanthropologist and solid science type, weighed in on the matter. A bona fide expert on foot structure in human beings and apes, he noted that the animal walked like a man, but with a strange cadence and exaggerated arm movements. Then he got specific, pointing out that the skull of the beast had a crest on top, like a male gorilla's, but unlike that of any female primate. Could this suggest that someone with not quite enough knowledge of apes and humans had designed this animal?

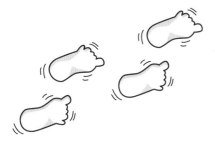

Napier wasn't finished. He argued that a Sasquatch's center of gravity would be higher than a human being's, and that would change its gait. But this animal walked like a human, and must therefore have a human-like center of gravity. If it looks like a duck and walks like a duck . . .

But it was the creature's buttocks that sealed the deal. Napier noted they didn't belong—at least they didn't match the upper body. Muscular buttocks are a human feature, so while Napier thought the upper half of the creature was ape-like, the lower half was human-like. That's impossible naturally, so one of those halves had to be artificial. He was pretty sure it was the upper half. But even as he raised doubt, Napier refused to shut the door totally on Sasquatch. He felt, partly based on the number of people who claimed to have seen it, that it was still impossible to say, "It does not exist."

Greg Long, however, was driven to put the Sasquatch story to rest forever. In 2004, he published a book called *The Making of Bigfoot: The Inside Story*, in which he claimed to have found both the man who wore the gorilla suit and the man who supplied the suit. He identified the suit wearer as Bob Hieronimus. Long claims to have taped Hieronimus both walking normally and walking as a Bigfoot and put that tape side by side with the Patterson film. The two matched. Admittedly, there are inconsistencies in Long's account: Hieronimus remembers the suit being a three-piece affair, while the man who purportedly made it claims it had six pieces. Why the breasts? Apparently, they didn't come with the suit. Why did Hieronimus remember the suit having a terrible smell when it was an off-the-shelf gorilla suit? Why do Hieronimus's legs and arms not correlate with those of the animal on film? And what happened to the suit? All good questions with no answers.

But the film footage of Sasquatch is only one piece of so-called evidence. What about the footprints? Sasquatch footprints are scattered across the Pacific Northwest. Over a hundred thousand have been found in the past fifty years, most of them obviously faked but a fraction that are harder to explain away.

I'll take an idealized print, one of the more believable samples, and start there. First, there's no arch. There is no arch because the tendons and ligaments that maintain the arch would be unable to support the weight of this animal. Second, the big toe is no larger than the rest. We humans stride off our big toe, but we don't weigh as much as a Sasquatch. And third, the heel leaves a deep imprint on the inside of the foot, not the outside, a reflection of the animal's weight. All these differences could be consistent with a non-human, ape-like form.

John Napier was convinced there were two kinds of Sasquatch footprints, implying two different species. He declared that one had to be fake. At the same time, he maintained there was "a curious and persuasive consistency" about the footprints, especially the variety he considered natural: the so-called Bossburg prints.

Found in 1969 near Bossburg, Washington, these prints formed a trail of 1,089 impressions. They were huge—17 1/2 inches (45 centimeters) long and 7 inches (18 centimeters) wide—but more significant, this Sasquatch had a clubfoot. One footprint was normal, but the other curled into the shape of a parenthesis, with what looked like calluses on the outside of the foot.

Experts argued about whether this reflected an inherited condition or was caused by an accident.

Human beings born with this congenital *talipes equinovarus*, colloquially called clubfoot, often rest only the front part of their disfigured foot on the ground. But these prints clearly showed the heel. Some felt the unusual footprint was therefore the result of a childhood injury.

 Are you as far down the rabbit hole as I am? These are anatomically correct clubfooted Sasquatch prints. Who would fake those? Even better, who would be qualified to fake those? Sasquatch believers claim a faker with such skills would have to be the equivalent of an artist/scientist like Leonardo da Vinci, thereby proving the beast is real. Sasquatch skeptics point out that books full of illustrations of disfigured feet exactly matching the Sasquatch print can be found at the Washington State University libraries at Spokane, two hours away.

If there's no Sasquatch—and I don't see the chances getting any better—then every footprint is a fake. And that means that many individuals from Alaska to California have, directly or indirectly, well or poorly, conspired to create a mass footprinting of the land. It's hard to imagine, but it's even harder to think the creature has eluded us all these years.

There are always those who pursue and perpetuate supernatural mysteries with incredible fervor—they are often more fascinating than the mysteries themselves. For the record, I lost interest in Sasquatch when one was spotted in a mall in Wisconsin.

History Mystery

Did Archimedes really set Roman ships on fire with the sun?

THIS STORY ISN'T AS WELL KNOWN as the one about Galileo dropping balls from the Tower of Pisa or Newton sitting under the apple tree, but it is more dramatic, more fabulous and even more contentious.

In 213 BC, Rome was at war with Carthage, the seafaring nation operating out of what is now Tunisia. The island of Sicily was largely occupied by Rome, but the city of Syracuse on the island was controlled by Carthage. The Roman army and navy attacked the city, but it was well fortified with a walled harbor, and the Greek genius Archimedes was overseeing its military. He was undoubtedly one of the greatest scientist-mathematician-engineers of the time, if not of all time.

Archimedes's exploits in 213 BC were more spectacular. The Romans had giant oared ships armed with devices called sambucae—huge ladders with tons of rock counterweights at the bottom. These could be swung up to rest against the harbor's walls, allowing the soldiers to scramble over the ramparts and enter the city.

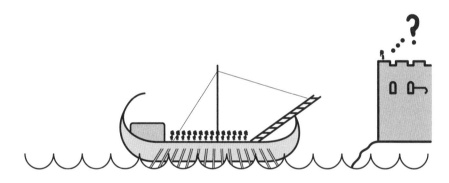

If the Romans thought sambucae gave them the technological edge, they were sorely mistaken. Archimedes had set up a variety of devices to ward off the ships, the most remarkable of which was (if the story is true) a giant mirror (or set of mirrors) that reflected the sun's rays onto the Roman ships and set them on fire. Imagine the terror and confusion: one minute you're getting ready to breach the walls of a city, the next your ship is ablaze.

How could Archimedes have done it? A number of experts have tossed their ideas into the ring.

First, with a mirror of the correct shape, it is physically possible to concentrate the sun's rays. But in this case, a single mirror would likely have been much too large and unwieldy, given that the Roman ships were at least a bowshot from the shore. A set of small, square mirrors, however, each held at exactly the right angle, would have behaved like one large mirror.

If Archimedes had arranged a tight formation of soldiers on shore, each holding his own mirror, a beam of sunlight hot enough to ignite wood could have been created.

But how would each soldier know exactly where to aim his mirror? In 1973, a physicist named Albert Claus came up with a theory: maybe each soldier had a two-sided mirror with a tiny hole cut in the middle. If each soldier aimed the mirror so that the Roman ship was visible through the hole, the spot of light from the hole would fall on the soldier's cheek. The soldier would see that spot in the reflection on his side of the mirror. If he adjusted the mirror so that the spot disappeared into the hole, he would be aiming the light directly at the ship.

Now imagine hundreds of soldiers all doing the same thing. That would make for one seriously hot weapon—although a recent analysis suggests it would have taken 420 soldiers, each holding a mirror the size of a card table, to reflect enough spring sunlight to light a Roman ship. Or would it?

Many people have tried using mirrors to light things on fire. The most notable pyromaniac was the great scientist Comte de Buffon, who in 1747 arranged eight-by-six-inch (twenty-by-fifteen centimeter) mirrors and focused them at a target of wood smeared with tar. On a clear spring day in Paris, 128 of these mirrors caused a piece of wood 150 feet (45 meters) away to burst into flames—smoke and mirrors indeed. Buffon tried other distances and numbers of mirrors and was successful at igniting a fantastic range of items.

Science _Fact!_ In 2013, the owner of a Jaguar in London, England, returned to his parked car to find that the side mirror and panels had been damaged; the car had partially buckled, and there was a smell of burning plastic. As it turned out, the glass on the side of a concave building nearby had focused the sun's rays on the car. The building has since been furnished with shades. The Jaguar has been repaired.

In the early 1970s, a group of Greek sailors standing on a beach holding bronze-coated mirrors set a plywood silhouette of a Roman ship on fire from about 165 feet (50 meters) away. But more recent tests, including ones by MythBusters and MIT, produced less exciting igniting. The MIT team managed to set fire to a model ship after about ten minutes. MIT and MythBusters together generated lots of charring but very little flame.

While these experiments have shown that wood can be ignited by mirrors, the circumstances have to be precisely orchestrated to ensure success. And it's very unlikely that would have happened in the harbor at Syracuse. After all, the Roman ships would have been tossing and turning in the wind, making it almost impossible to focus a sunbeam on a particular spot long enough to cause ignition. Plus, the wood of the hulls would have been wet, and clouds would have dramatically diminished the power of the beam being reflected.

Given all of that, and given that Archimedes had already invented ingenious war machines—like a giant claw that could hook the prow of a Roman ship and tip it on end—it's unlikely that he would have spent a lot of time trying to make such a fallible, unpredictable technology as sun and mirrors work.

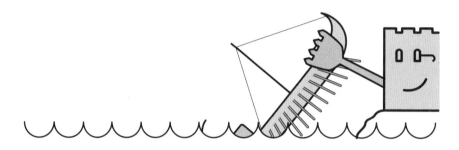

Also, it's not even clear that Archimedes ever had the mirror idea in the first place. The historian Polybius, who lived at the time of the battle, described a number of technologies that Archimedes invented and deployed against the Romans, but mirrors were not among them. Two other contemporaries who wrote about the siege also failed to mention mirrors.

A mathematician named Diocles wrote about "burning mirrors" in the decades after 213 BC, and while he did mention Archimedes in a mathematical context, he made absolutely no mention of mirrors in the siege of Syracuse. What's more, Diocles claimed he was the first to prove that a parabolic mirror could concentrate the sun's rays.

In the face of this tidal wave of uncertainty, one tiny refuge is that mirrors could have been used to momentarily startle and blind sailors. That would have given the Syracusans enough time to deploy some of Archimedes's other fine weapons of war.

Science Fiction! *Science fiction author Arthur C. Clarke wrote a story called "A Slight Case of Sunstroke," set in the fictional country of Perivia in South America. Half the seats in a stadium are occupied by soldiers with souvenir programs silvered on the back. After a particularly bad call by the referee, the soldiers deploy their mirrored programs, focus the sun and incinerate him on the spot. Not likely, but entertaining!*

Part 4
The Natural World

.

Can humans echolocate?

THE SHORT ANSWER IS YES, and there are individuals who prove it. Daniel Kish has been blind since the age of thirteen months, but now, as an adult, he can camp in the woods, swim, dance and otherwise navigate life, all thanks to his extraordinary echolocation skills. Other blind echolocators can navigate around a city by skateboard or bicycle. And yet, even with this evidence staring us in the face, it seems unbelievable that humans can echolocate. Most of us have never had the slightest hint that we might be able to do so. But if you pause for a moment, you can see many everyday examples of echolocation in practice—a contractor knocking on a wall to determine where the studs are, or a doctor tapping a patient's abdomen to assess the health of the organs within.

This general lack of awareness is the reason that the first investigators who explored the mechanics of echolocation came up with some weird explanations. Some believed that echolocation was made possible by changes in air

pressure on the eardrums or the skin of the face (an early term for it was "facial vision"). Others attributed the cause to magnetism, electricity or the ether. One expert even argued that touch receptors in the skin were actually tiny eyes, and at one point, the subconscious was thought to be the source of the ability. Although those ideas were inventive, they lacked the benefit of experimental evidence. Finally, around the turn of the twentieth century, Theodor Heller, a German scientist, actually tested blind people's abilities to sense the presence of an obstacle in front of them. He declared that his subjects were able to pick up changes in the sounds of their footsteps from nine to twelve feet (three to four meters) away from the obstruction, and so, he concluded, acoustics provided the crucial signals that allowed people to navigate without seeing.

The first echolocation experiments were conducted at Cornell University in the early 1940s. The researchers had their subjects—two blind and two sighted—walk down a long hallway that was either empty or had a large sheet of Masonite randomly placed in it. To make the experi- ence as consistent as possible, each subject was blindfolded, and the experiments were run either on a bare wooden floor with the subjects wearing shoes or on a carpet with the subjects wearing socks. When the subjects felt that they were approaching either the end wall or the Masonite board, they were to raise their right hand when they first became aware of the obstacle and then raise their left hand when they felt they were about to bump into it.

Striking results emerged from these tests. As you'd expect, the blind participants fared better than the sighted ones at first: the sighted subjects ran into the wall more than a dozen times during the first trials and often veered so far sideways that they risked running into the side walls of the hall. The top performer initially was one of the blind subjects who was consistently able to detect the wall from more than twenty feet (six meters) away. When the testers asked him how he could sense the obstacles, he explained that somehow the sensitivity of his face was allowing him to do so. He was so convinced of this that he said the wall cast a "shadow" on his forehead that he could feel, even if he couldn't see it. He dismissed the idea that he was actually using his hearing.

The subject's explanation went against the scientists' expectations, and so the Cornell team began to refine the experiment to test how exactly their top performer did so well. After the subject performed dramatically more poorly while wearing socks on the carpet, the scientists realized that it was his ability to hear his footfalls, not some phantom presence felt on his face, that was the key to his success. To ensure that sound, not touch, was the important factor, the testers ran another version of the experiment in which they enclosed the man's arms and hands in leather gauntlets and covered his head with a hood that extended over his chest.

With all that equipment on, the subject could still hear, but he was unable to detect a breeze from an electric fan placed ten feet (three meters) away. Even outfitted like that, he could still sense the presence of an object in his path.

With that new information in hand, the scientists ran the tests with all the subjects wearing earplugs. They all performed less well, proving that hearing was crucial to echolocation, for both blind and sighted subjects. The more trials the scientists ran and the more they refined the experiment, the better the sighted group became at echolocation. By the end, one of the sighted subjects actually performed better than one of the two blind ones.

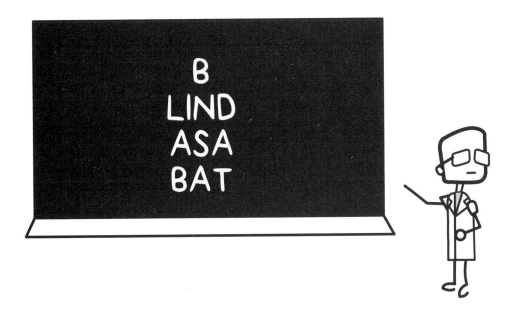

These initial experiments were followed up by variations of one kind or another in various labs, all of which supported the idea that echolocation is a hearing sense, not an air pressure sense. Later experiments also started to flesh out the details of this hearing sense, such as the fact that echolocation relies on the pitch of the sound issued, not its volume. In one study, subjects were encouraged to make whatever noises they felt would best reveal their environment. Instead of just relying on their footsteps, the subjects made clicking noises, snapped their fingers, hissed, whistled and even sang the "do, re, mi" scale. The common element of the most effective strategies was that the subjects exploited the higher-frequency sound range.

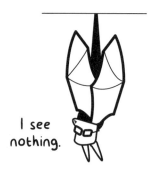

I see nothing.

Another study showed that echolocation was more successful when the subjects were able to move their heads from side to side. Subjects were trained to echolocate their way down a hall, then they tried to do the same in a video simulation of the same hall, using the same echolocation sounds. In one version they were allowed to move their heads and bodies to align themselves with the center of the hall; in the other, they had to remain still and adjust their heads with a joystick. The joystick part of the experiment was a failure—the subjects consistently ran into the walls, showing that head movements are crucial to being a good echolocator, probably because of the slight differences in patterns of echoes depending on where the ears are pointing

 TRY THIS: Hold your hand, palm facing you, at arm's length in front of your face. Begin hissing. Slowly move your hand closer to your mouth. You will notice differences in the sound. It changes as a result of the outgoing hiss interfering with the returning echo. It might be difficult for you to attribute this directly to the echoes (just as it was difficult for one of the subjects in the Cornell experiments to believe he was using sound and not pressure on his face), but there's no mistaking the fact that something is changing.

One of the most fascinating discoveries about echolocation has been the modern revelation of what goes on in the brain while it's happening. Mel Goodale and his colleagues at the University of Western Ontario performed an experiment in which they placed two adept echolocators—blind individuals who were able to play basketball or to mountain bike—into an MRI and mapped the areas of their brains that were activated by echoes. They found that the activity in auditory parts of the subjects' brains did not ramp up when hearing echoes, but the activity in the parts of the brain responsible for visual information did. As these individuals began to rely increasingly on echolocation for their perception of the world around them, they apparently abandoned the auditory parts of the brain and recruited the visual. In the case of these particular individuals, the one who had become blind more recently did not show as strong a response in the visual areas of the brain, suggesting that it takes time to reroute this information.

You'd be hard-pressed to match a subtle and sophisticated piece of biology like this with a machine. What's more, humans have proven themselves able to train and improve that ability, suggesting that it's a skill within everyone's reach, regardless of their dependence on sight. What's more, those who echolocate are not limited to simply localizing objects, but can also make accurate judgments about shape and motion. Experienced human echolocators can determine the difference between two objects only 1 inch (2.5 centimeters) apart located 3 feet (1 meter) away, and some can even differentiate by echo the difference in texture between velvet and denim. Contrast that with the sad fact that sonar operators in submarines have often failed to discriminate between a whale and an enemy vessel and as a result depth-charged the whale. Most of us aren't aware we have these tools built into us, but that's simply because we've never had any reason to think about them or to try them out.

Of course, no matter how talented some humans might be at echolocation, bats are the gold standard in the field (closely followed by dolphins and whales). The discoveries that led to our recognition of bats' mastery of biological sonar were made by a curious collection of people. First, Lazzaro Spallanzani—better known for being the first person to perform in vitro fertilization (in frogs)—demonstrated in 1790 that bats can navigate without being able to see. Unfortunately, he proved that by putting out their eyes and showing that the loss of their eyesight wasn't a hindrance to their ability to fly.

Next, Hiram Maxim, the inventor of the machine gun, proposed that bats emit sounds below the range of human hearing and listen for the returning echoes. He didn't test his idea with experiments, but he was correct on one count—we can't hear bats' sounds—and wrong on another—their calls are actually above the range of human hearing. (After the *Titanic* sank in 1912, Maxim suggested that ships could use echolocation to detect icebergs at long range.)

It was eventually established that bats use ultra-high-frequency sounds to detect and capture their insect prey. That sounds easy, but it most definitely is not. Consider that a bat hunting near trees, bushes and buildings must be able to distinguish echoes returning from those objects. And it must be able to distinguish this noise "clutter" from the wingbeats of a flying insect. Bats do that by changing the frequency of their clicks. When a bat is scanning its environment, it emits about ten to twenty clicks every second. But when it picks up an echo returned from a target, it increases that rate to as many as two hundred clicks a second in order to zero in on the position and identity of its prey.

Some species of moth, upon hearing a bat's clicks, will instantly fold their wings and drop straight down to avoid getting caught. Bats, in turn, have evolved ways of adjusting their clicks to take in a wide area in which even a plummeting bug will be tracked. Some moths have counter-adapted and evolved long extensions to the backs of their wings that twirl in flight to attract bats' attention. The extensions are disposable to the moth, so even if a bat succeeds at biting them, the moth will simply break away and escape. It is a never-ending battle between predator and prey—one in which bats will surely adapt once again to refine their already stellar ability to echolocate. But let's see one ride a bike while doing that.

Why does time seem to speed up as we age?

THERE'S NO DOUBT THAT THE VAST MAJORITY of people feel that time moves faster as they age, but very few of them bother to estimate by how much.

A century ago the great American psychologist William James suggested that as we grow older, and more jaded and worldly, we enjoy fewer remarkable experiences in a year, and so the years become less and less distinct from each other. Another theory suggests that because each successive year is a smaller percentage of one's overall life, it is less significant when weighed against the rest and therefore passes by virtually unnoticed. When you were ten, every year was huge: 10 percent of your life. At age forty, though, one year is only 2.5 percent of your total life.

Sometimes I feel time is catching up to me.

There's also a phenomenon called forward telescoping. Imagine you're asked when you last saw your aunt and you say, "Uh . . . three years ago?" when it's actually eight years since you saw her. You've zoomed in time, bringing the past closer than it really is. When someone asks me how long ago an event took place, I double my first estimate, and even then I sometimes underestimate the passage of time. That's forward telescoping.

In the mid-1970s (remember how slowly time passed then?), Robert Lemlich of the University of Cincinnati proposed one significant adjustment to the idea of the apparent passage of time versus reality. He argued that since time is all subjective anyway, years are also subjective. Calculating what percentage of your total life is represented by each passing year is fine, but it's strictly mathematical and so doesn't take into account that each passing year feels shorter as well—it is a smaller part of your total life numerically, but it feels even less than that. It's all in your head, really, so your estimate of the length of a year that has just passed should be compared not to how long you've lived but to your sense of how long you've lived.

Lemlich created equations to quantify what he meant. Their implications are surprising, even shocking. Let's assume you are a forty-year-old. Lemlich calculated that time would seem to be passing by twice as fast now as it did when you were ten. (Remember how long summer vacation seemed to last?)

But there's more: the numbers tell you that if you're that forty-year-old and you're going to live to eighty, you're halfway through your life by the calendar, but because time seems to be passing ever more rapidly, Lemlich's math suggests you will feel you have less time left than you actually do. By his calculations, at age forty, you have already lived—subjectively—71 percent of your life. It gets worse: by the time you're sixty, even though you have twenty years remaining, those twenty years will feel like a mere 13 percent of your life.

These numbers are shocking enough, but they take on an even more bizarre twist when you extrapolate them back and ask the question: At what point in our lives have we experienced half of our subjective life? If you're that forty-year-old, you will have experienced half your total subjective life by the time you were twenty. Even if you live to a hundred, 50 percent of your total life experience will feel locked in by your twentieth birthday.

Lemlich backed up his numbers with experiments. He asked a group of students and adults to estimate how much slower time seemed to have passed when they were either half or one-quarter their present age. His theory predicted the answers almost exactly: time seemed to have passed only half as fast when they were one-quarter their present age, and about two-thirds as fast when they were half their present age.

Is something else going on in our brains that would change our perception of the passage of time as we age? It might be that our internal clock (and jet lag and shift work demonstrate just how crucial that clock is) runs slower as we age. If your clock now estimates a minute to be three minutes, because it's running slower, then many more events will be packed into that time frame and it will seem that time is passing faster.

An extreme example is the case of a man who, at the age of sixty-six, was admitted to hospital in Düsseldorf. Examination revealed a tumor in the left frontal lobe of his brain. He'd gone to the hospital because he was finding life unbearable: everything was happening at breakneck speed. He had to stop his car by the side of the road because the traffic was too fast. The television, already manic, was triple-manic, and as a result of this experience, he had begun to withdraw from society. When asked to estimate the passage of sixty seconds, it took him four and a half minutes. Imagine what traffic would look like if four minutes' worth was packed into a minute! What this case suggests is that disruptions

How to stop time

to certain parts of the brain alter our perception of the passing of time, and while this particular case was unusual, it's possible that a gradual and minor version of this affects everyone's sense of time passing.

 TRY THIS: Want to get a sense of your subjective age? It's simple: Estimate how old you will be at death. Divide that number into your current age. That gives you the percentage you've lived of your chrono-logical age. But if you then take the square root of that, you have the percentage you've lived of your subjective age. Be ready to be shocked! Here are my numbers. I come from long-lived parents, so I should live to ninety. That's twenty more years—yay! But that twenty represents only 11 percent of my entire life experience. That means that I have already experienced close to 90 percent of my subjective life. Boo.

You might be wondering why we're spending time (it's precious!) figuring out equations to account for how we experience time. This kind of data supports what might otherwise seem to be mere impressions like this one by Robert Southey, the poet laureate of England in 1837: "Live as long as you may, the first twenty years are the longest half of your life. They appear so while they are passing; they seem to have been so when we look back on them; and they take up more room in our memory than all the years that succeed them."

 TRY THIS: The next time January rolls around, pay attention to the number of times you date anything and write the previous year. I used to do that consistently, almost to the end of January, but in the last two years I haven't once written the incorrect year. I have no idea why. Keep an eye on it yourself. Report back to me what you learn!

Why does campfire smoke follow you around?

MANY PEOPLE HAVE ARGUED that when you move around a campfire, you leave a partial vacuum behind that sucks smoke toward you. But nature truly abhors a vacuum, and creating even a momentary partial vacuum is extraordinarily difficult. So how does smoke seem to always know where you are?

The real answer has to do with airflow. Typically, cool air from beyond the perimeter of a fire flows toward the flames, warms and then rises. As long as there is no disturbance in the air, there is an even and steady airflow. But if you change position, or even worse, stand up and start moving around, you disturb that air cycle, creating an eddy in front of you. This always happens when a fluid (in this case, the air) meets an obstacle (your body). It's not unlike the effect of a rock in the middle of a stream or dead leaves scattered in the wake of a bus.

Let's say you're facing the fire. The incoming air reaches your back and is channeled to either side of you as it flows past. But the two streams don't stick to your sides and immediately rejoin at your belly button. They flow straight past, leaving an empty space in front of you that needs to be filled. As the two streams of air near the fire, they reverse direction, flowing back toward you. As they move back toward you, they carry the smoke from the fire with them. The next thing you know, you're enveloped in a cloud of campfire smoke. Not only that, but the bigger you are, the more smoke gets drawn toward you—the bigger the obstacle, the more air gets displaced and pulled back into the eddy. Change places and you'll simply trigger the same effect wherever you go.

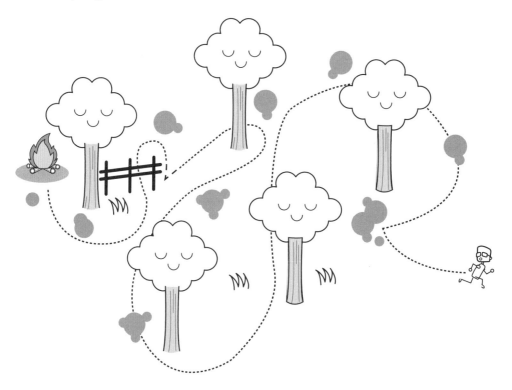

Another thing that might affect how smoke follows you is the design of the fire itself. Adrian Bejan of Duke University has calculated that the ideal shape for the logs in a fire is a cone or pyramid, where the height is equal to the length of the base. Bejan argues that a cone perfectly balances airflow and geometry to generate the most heat for those sitting around the fire; basically, the fire gets hotter because it can breathe better.

But there's another benefit to this geometrically ideal campfire. Its height and shape allow a smooth smoke plume to form and rise more or less straight up from the core of the fire. If the smoke rises from the center, it is less affected or disturbed by motion on the edges of the fire pit, and so it's less likely to surround you as you move.

Of course, that's all presuming that you're standing around the fire in ideal conditions. Once you throw in wind, weather and other people moving around, you have a perfect storm of physics in which all sorts of things can happen. Some campers advise placing a large stone or stump—one that surpasses you as a source of turbulence—on one side of the fire to act as a permanent obstacle. Bejan's reaction to that idea is that it had better be a big enough stone—in his words, "an obelisk." Or you could just make sure you're not the biggest person standing at the fire.

Is it a small world after all?

It depends on how you define "small world," but if you're referring to the idea of "six degrees of separation," then yes, we are indeed in cramped quarters. Although the evidence is mixed, any one person can connect herself to anyone else in the world by six steps or fewer.

The phrase "six degrees of separation" has a long history. A Hungarian author, Frigyes Karinthy, coined it in a short story he wrote in 1929. A character places a bet that he can connect to any one of (what was then) the earth's 1.5 billion people using no more than five intermediaries, the first of which would be a personal connection. Most people today are less acquainted with Karinthy's short story and more familiar with John Guare's play *Six Degrees of Separation*, or at least with the subsequent film of the same name.

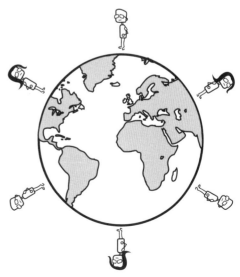

The "six degrees" idea was first explored scientifically in "The Small World Problem," a 1967 study authored by Harvard psychologists Stanley Milgram and Jeffrey Travers. Milgram (better known—or more notorious—for his obedience experiments) was given $680 to test the idea that any two Americans could be connected by a relatively short series of acquaintances. Milgram chose Omaha, Nebraska, and Wichita, Kansas, as starting points for his test, and the Boston area as the finish line. Volunteers in both Omaha and Wichita were tasked with connecting to one of two strangers—the wife of a divinity school student in Cambridge, Massachusetts, or a stockbroker in Boston. The volunteers in Omaha and Wichita were asked to pass on a letter—by hand or by snail mail—to someone they knew on a first-name basis who might help send the package closer to its final recipient.

Did You Know . . . The six in "six degrees of separation" refers to the number of links in the chain. So seven people are connected by six links, with the two people at each end of the chain connected by five people between them.

Milgram and Travers were surprised to find that the first successful connection arrived in four days and had required just four people (and three links) from beginning to end: a wheat farmer, a local minister, an instructor at the Episcopal Theological School in Cambridge and the wife of the divinity school student. As other results began to trickle in, they found the number of contacts needed to complete the transaction ranged from two to ten, with five being the median.

The numbers obscured the reality, though: the volunteers weren't random enough. In the case of the man in Boston, a number of stockbrokers in Nebraska were chosen to forward the document to him, offering an advantage through professional association. And when it came to the final handoff to the target, the last person in the chain was frequently the same gatekeeper. For instance, the Boston stockbroker was reached both at home and at his office, but of the twenty-four transfers that reached him at home, sixteen were delivered by the same man.

A final problem was that the sample sizes and numbers in the experiment were quite small: of fifty letters sent out, only three ended up on the doorstep of the wife of the divinity student, and the average number of steps in those successful deliveries was eight. The stockbroker chain performed better, but still, fewer than a third of the original 160 letters completed their trips. The expeirment's widespread fame, even outside the scientific community, obscured the fact that the data were weak.

From the 1970s through the 1990s, few social scientists replicated the original experiment. But the era of social media brought a whole new set of resources for figuring out the exact number

of degrees of separation among us. In 2007, a study using Microsoft's instant messaging system reported that among the 240,000,000 people who had generated thirty billion conversations, the average degree of separation was 6.6. It was claimed that this was the first time a "planetary-scale social network" had been analyzed.

Then, in February 2016, Facebook released an analysis of its 1.59 billion users. They found that no one was any more than three and a half people away from anyone else—3.57 degrees of separation, to be exact. By that count, you would need the connections supplied by three and a half other people to get to anyone else on the planet. Of course, there are still five billion people who aren't on Facebook. Whatever the exact number of degrees of separation, it's shrinking as our networks grow.

While the computer, sociology and psychology worlds have been modeling, experimenting and debating the six degrees question, pop culture has embraced it as a given. Just check out the Oracle of Bacon website if you don't believe me. The site theorizes that Kevin Bacon has appeared in so many films that he's separated by less than six degrees from any other living actor. And considering that Stan Lee and his Marvel Universe are everywhere, it isn't much of a surprise that a "small world" experiment has been done with them, too.

In 2002, authors Ricardo Alberich, Josep Miro-Julia and Francesc Rosselló created a network that mapped superheroes to each other. They found that, on average, each Marvel character had appeared in fifteen comics. Spider-Man had the most appearances, at 1,625, but Captain America had the most connections, at 193. Given the Captain's range, he needed only 1.7 other characters to be within reach of everyone in the entire Marvel Universe. There weren't nearly as many clusters of related people (related through profession, for instance) in the Marvel version as you'd expect in a real-life social network of that size. In reality, we cluster into overlapping communities of friends and colleagues. Superheroes do not. Too much to do, I guess.

Super Science Man

Does water drain from the bathtub in two directions?

THIS QUESTION CAN BE ANSWERED in two different ways: How does water theoretically drain versus how does it actually leave the tub? Most of the time, we want to know how things actually work, but in this case, the theory is way more interesting, even if it is more difficult to demonstrate.

As water swirls down the drain, all things being equal (and you'll soon see just how critical that caveat is), it will swirl counterclockwise in the northern hemisphere and clockwise in the southern half of the world. The difference is best visualized by picturing a TV weather map. If a major storm or hurricane in the northern hemisphere appears on screen, you see the massive, nearly circular counterclockwise swirl of clouds. The wind and clouds' movement is the result of air flowing from a zone of high pressure to one of low pressure. That seems relatively linear, so why don't the clouds move in a straight line? Blame the earth that's turning beneath them.

The earth rotates from west to east, and depending on where you are on the planet, you are being carried along by its rotation at a unique velocity. At the equator, that velocity is 1,038 miles an hour (1,670 kilometers an hour). At the North Pole, it's zero. In Los Angeles, you're turning eastward at about 857 mph (1,380 km/h); in Toronto, you're moving at roughly 745 mph (1,200 km/h); and in Yellowknife, you're cruising at 479 mph (770 km/h). With that in mind, if you're in an airplane in Los Angeles and you plot a straight-line course to Yellowknife, you won't even come close to the Northwest Territories capital. You would begin your journey with an eastward velocity of 857 mph (1,380 km/h), which means that right from the get-go you'd be rotating to the east nearly 435 mph (700 km/h) faster than Yellowknife, and so you'd end up far to the east of your destination. To make things easy, let's say you would end up to the right. And if you tried the same thing on the return flight, you'd finish far to the right (west) of the City of Angels, because the earth's rotation would have moved the city eastward faster than it was moving you the same direction. It would all be the opposite in the southern hemisphere, of course, where trying to take a straight-line flight would have you end up to the left (or west) of your target.

The same effect explains the northern hemisphere storm winds. From the clouds' point of view, they're flowing straight toward a center of low pressure. But the earth is turning beneath them, and so they're constantly missing just to the right. The result is that they travel in ever-decreasing circles, each bending in whatever direction their hemisphere dictates: hurricanes in the North Atlantic swirl counterclockwise, whereas typhoons in the South Pacific swirl clockwise.

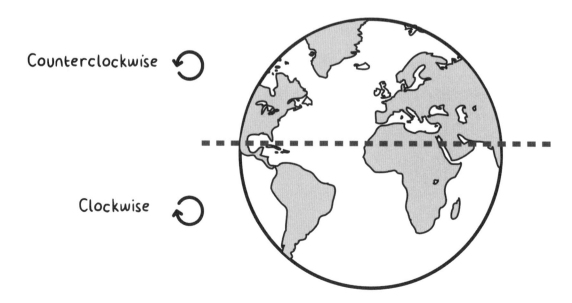

Counterclockwise

Clockwise

The name for this phenomenon is the Coriolis effect, and it isn't limited to storms. In fact, anything traveling across the surface of the earth is affected by it. In World War I, the Germans deployed a huge piece of artillery called the Paris Gun. It fired a shell into the stratosphere that could travel up to 80 miles (130 kilometers), but the Coriolis effect displaced the shells about 1 mile (1.6 kilometers) to the right of their targets. Although the effect wouldn't have been nearly as obvious, even the arrow from William Tell's crossbow flying 120 paces to split the apple balanced on his son's head would have deviated ever so slightly to the right because of the Coriolis effect.

The Coriolis effect touches everything, but it's more obvious when the moving object is large and traveling a long distance. For instance, a baseball traveling on a 500-foot (152 meter) home run would be affected by the rotation of the earth, but the effect would be so minuscule that it could be detected by only the most elaborate laser-ranging equipment available. Compare that to the plane flight from L.A. to Yellowknife, where you'd miss by miles.

In theory, then, even water draining out of the bathtub should be influenced by the Coriolis effect. The water should drain counterclockwise in the northern hemisphere (always missing the drain to the right) and clockwise in the southern; if the water is the clouds in our weather example, then the drain plays the role of the center of low pressure.

The problem is that on the scale of a bathtub, the Coriolis effect is so slight that any number of other influences will overwhelm it. The mere act of pulling the plug, small currents or flows made from someone getting out of the tub, the shape of the basin, a drip from the tap: all these will exert much more force than—and mask—any effect due to the Coriolis effect. The result is that even if the water happens to drain out of the bathtub counterclockwise in the northern hemisphere, you could never establish that the Coriolis effect was solely responsible.

There have been countless demonstrations that have tried to remove these everyday obstructions and accurately show the Coriolis effect in action. My favorite is a classic pair of tests held at MIT and in Sydney, Australia, in the 1960s. Both teams of researchers went to great lengths to eliminate any disturbances that might have affected their experiment. They started by each building a tank 6 feet (1.8 meters) in diameter and placing a drain flush with the center of the bottom of the tank. The Australians even went so far as to make their tub out of wood to avoid any heat effects metal might have had, and they isolated their tank in a cement-walled, windowless basement room with restricted access. The teams then filled their tanks with water, careful to pour it into the tubs in the opposite direction of the way they expected it to drain; they wanted to eliminate the possibility that trace momentum left by filling the tank could influence the results. Both teams then covered the tanks and let the water sit for eighteen to twenty-four hours. When everything was finally set, they both pulled the plug out from underneath their tanks in a way that wouldn't disturb the water's surface. The result? It worked! Enough care had been taken that the results could be considered fair evidence that the Coriolis effect can—under very special circumstances—dictate how the water drains from the bathtub.

In theory, the Coriolis effect should disappear if you are draining the tub exactly at the equator. But again, to test this theory—to see the flow disappear down the drain without any observable rotation—you'd have to construct an experimental apparatus at least as elaborate as the ones from the 1960s. Clockwise, counterclockwise, no rotation at all, it doesn't matter—in an everyday setting, you can never identify the cause. However, at least you know what should happen, unlike Scully and Mulder in a 1995 episode from the first *X-Files*. As the pair investigates a series of mysterious events in a small town, Mulder notices that the water is draining counterclockwise, and he (incorrectly) points out that it should flow clockwise because they're in the northern hemisphere. Based on that false start, the detectives soon conclude there's some supernatural power nearby that's bending the laws of physics. Of course they do.

Clockwise or not...

...everything is going down the drain.

Why do leaves change color in the fall?

Considering how beautiful fall colors can be, you might think that it takes a lot of energy for leaves to change their hue from their summer green to the reds, yellows and oranges of autumn. But in fact the opposite is true: many of the color changes you see in the fall are the unintended byproducts of a cost-saving measure by trees.

Leaves are green because they contain chlorophyll, a molecule that collects solar energy. Chlorophyll absorbs all light across the sunlight spectrum except for green, which it reflects back to our eyes. During the growing season, when there are the most hours of daylight, chlorophyll is at peak performance, converting the energy of sunlight into sugar molecules that can release the energy later or be used to build other molecules needed by the

tree. But when sunlight diminishes in early fall (it shrinks steadily from June 21 through to December 21), the cost-benefit equation for the tree changes. The energy that a leaf absorbs from the sun dwindles to less than the energy it uses to perform the chemical work required to maintain its sophisticated machinery. To prevent wasting resources, the tree begins to shut down its leaves.

As the leaves shut down, chlorophyll molecules are shipped back into the tree. (Nitrogen is a limited resource, too, so the tree also stores it for use the next spring.) As the chlorophyll is withdrawn, other compounds in the leaf are left behind. Most of those compounds are carotenoids, and two of the most important are carotene and xanthophyll. Carotene is orange (carrots), while xanthophyll is yellow (squash). As the chlorophyll depletes and the carotene and xanthophyll are exposed, the color of the leaf changes.

That covers the oranges and yellows of fall trees, but what explains the fiery reds of maples that so delight tourists in eastern North America? The red in leaves can be attributed to a group of chemicals called anthocyanins, which, depending on their acidity, create a range of color from red to purple. Anthocyanins are the same molecules that give beets, grapes and plums their hue. The mystery is: What are anthocyanins doing there?

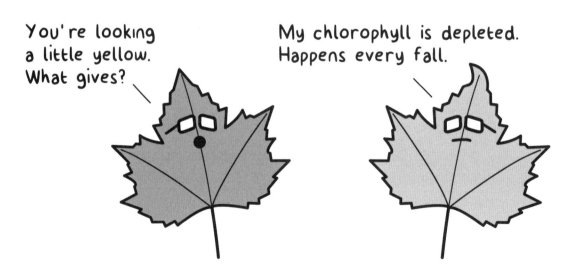

In the sunny days of summer, anthocyanins aren't present in leaves. But when the leaves' adjustment to winter begins, the tree manufactures them. Why on earth, at a time of depleted resources, would trees go to the trouble and expense of synthesizing new chemicals? At least the others, the carotene and xanthophyll, were there already. That is the mystery of the anthocyanins, and it has generated some interesting conjectures.

One is that the anthocyanin pigments protect against sunlight. At first, that sounds completely counterintuitive. After all, gathering sunlight is what leaves are all about, and on top of that, the change in color is happening at a time of year when sunlight is in decline. But there are circumstances when too much light can overwhelm the photosynthetic apparatus and actually damage a tree's leaves. This is more likely to happen during times of inadequate nutrition and low temperatures—such as autumn—when leaves are already dismantling their internal machinery. One unusually bright day in fall could cause significant damage to a tree, but thanks to the anthocyanins in its leaves, the tree can absorb high levels of sunlight.

Some circumstantial evidence supports this idea. For one, sun-facing sides of leaves are usually redder. The leaves on the south-facing side of the tree are redder too. And the farther north you travel, the redder the maples are—the higher the latitude, the cooler the nights, and so the more stressed the trees. If these observations are true, it suggests that trees do an incredibly delicate balancing act between getting enough light and getting too much.

The other suggestion as to why trees produce anthocyanins has attracted much more interest among scientists, likely because it lies at the heart of a fascinating theory of biological "gaming," the mechanisms and strategies by which organisms compete. Some people have proposed the idea that leaves turn red to protect themselves against insects that might lay eggs on them. For instance, in the fall, aphids lay eggs in crevices in tree bark, and these hatch after winter is over. Some scientists speculate that trees capable of manufacturing noxious, insect-inhibiting substances color their leaves red to advertise that fact to insects. Insects then lay their eggs elsewhere.

It appears aphids are in fact less attracted to red leaves than to green ones. But that doesn't mean their offspring fare worse on trees with red leaves. One experiment designed to sort this out compared wild and domesticated apple trees. Domesticated apple varieties are selected for the quality and abundance of their fruit, not for their insect resistance or leaf color. If red leaves are indeed a signal that aphids should back off, then you'd expect more red leaves in the wild varieties of apple trees, which aren't protected by farmers and pesticides. Conversely, if signaling isn't important, there should be no appreciable differences in leaf color or aphid activity between the cultivated and wild varieties.

The results of the experiment confirmed that trees with red leaves were more successful at warning aphids to stay away. The wild varieties of apple tree had a much higher percentage of red leaves in the fall than did their domestic counterparts. As well, the offspring of aphids that did lay their eggs on trees with red leaves had only about half the rate of survival of those from eggs laid on trees with yellow or green leaves.

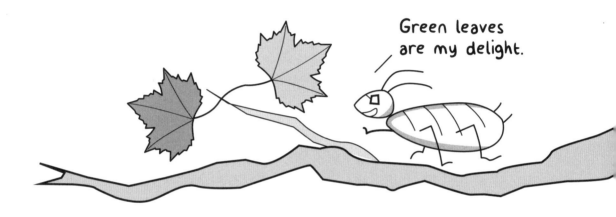

Sadly, one good experiment does not confirm a theory, and later discoveries put this particular idea on somewhat shaky ground. It turns out that aphids have no red receptors in their eyes. But maybe they are still able to discriminate between colors. And maybe it doesn't matter: as it turns out, it imight not be the red color itself that's warning insects away. Rather, the red covers up the yellow shades of the leaf (those exposed by the removal of the green chlorophyll). Those yellows are very attractive to a broad range of insects, including aphids. So the red hue isn't a stop sign after all. But because amber means "Go" to an aphid, disguising it with red has the same effect.

What is the sound barrier and what can break it?

IF YOU'VE HEARD AN ECHO, then you've heard evidence that sound travels at a particular speed. The speed of sound, at sea level, is about 760 miles an hour (1,225 kilometers an hour). A fast-moving object such as a jet plane can actually catch up to the speed of sound, in effect causing that sound to pile up in front of it. When it bursts through this barrier—the sound barrier—the pressure change is heard as a sonic boom.

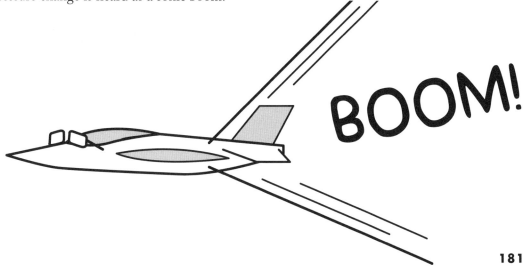

But planes aren't the only things that can break the sound barrier. Believe it or not, bullwhips can, too. When the tip of the cracking whip breaks the sound barrier, it creates a sudden, sharp noise.

How can a whip generate that kind of speed? It's all in the design. A bullwhip is thickest at the handle and gradually tapers toward the tip. When an expert cracks that whip, a wave starts from the handle and travels the length of the whip. Physics demands that the energy from the wave must be conserved. With the mass (the amount of material in the whip) gradually lessening as the wave moves toward the tip, there's more energy available to be converted into velocity, resulting in a wave that travels faster and faster. By the time the energy reaches the end of the whip, it is often going faster than the speed of sound—creating a loud sonic boom.

Science Fact! *One mystery that has plagued the understanding of whip cracks is that by the time you hear the crack, the tip has long passed—even doubled—the speed of sound. New evidence suggests that it is actually the loop that creates the crack rather than the tip of the whip. Either way, though, the sound is impressive.*

But bullwhips aren't the only common objects that make a cracking good sound. The poor man's version of the whip crack is the wet towel in the change room. There is no doubt that when properly flicked, the towel's sound is impressive, but is it a sonic boom? As is often the case with a totally unimportant question, there are enterprising students somewhere in the world who have tried to answer it. In this case, it's not just students but their teacher, too—at the North Carolina School of Science and Mathematics.

In 1993, they recorded up to ten thousand images per second of a flicked towel. It was a demanding setup, and they were rewarded by captured images of the tip of the towel clearly surpassing the speed of sound! To their credit, they knew this wasn't enough evidence to prove that the sound was a sonic boom, just that the towel had reached sufficient velocity to produce one. So they modified their experiment, using sound to trigger the photo-taking. They found that the highest velocity of the towel tip and the blast of sound coincided. Still, there were puzzles, because they also recorded a snap or two when the towel did not exceed the speed of sound. They reasoned that in these cases the period of time the towel broke the sound barrier was briefer than the intervals they were recording. When they reduced those intervals, they proved that indeed the sound barrier had been breached.

That was pretty cool work (and nobody appears to have replicated it since), but the absolute ultimate in flicking something to break the sound barrier takes us back something like 150,000,000 years, to the age of the dinosaurs—specifically, to the giant herbivore *Apatosaurus*. Yes, that's right: some dinosaurs might have been able to snap their tails so fast that they broke the sound barrier.

The *Apatosaurus* was huge: sixteen to seventeen tons (fifteen to sixteen tonnes)—three to four times the weight of an African elephant—and sixty-five feet (twenty meters) long. Just over half of the creature's total length was devoted to just the tail. The tail of *Apatosaurus* and a bullwhip have one thing in common: their shape. They're thick at the base and taper all the way down to the tip. In a bullwhip, the handle is about six hundred times thicker than the tip. In *Apatosaurus*, that taper was even more dramatic: from three feet (one meter) thick where it attached to the body to a tip that was thirteen hundred times smaller. Each vertebra in the tail was 6 percent smaller than the one before, and there were more than eighty of them. Amazingly, the first twenty or so vertebrae supported 97 percent of the total mass of the tail.

Some have suggested that the very long tail was a defensive weapon. Others posit that it might have been part of a mating display. Nathan Myhrvold, former chief technology officer at Microsoft, and Phil Currie, a paleontologist at the University of Alberta, were intrigued enough by the tail's peculiar size and shape to investigate further. They wondered: With a tail that unique and that long, if the *Apatosaurus* flicked it, would it create a bullwhip cracking sound?

Working with one of the best Apatosaurus skeletons and filling in details from other skeletons where necessary, they reconstructed the length, weight and flexibility of the tail. They concluded that it was indeed enough like a bullwhip to have generated a fast-moving wave along its length. There were other encouraging details, too, like the fact that the vertebrae toward the tip had gaps between them, suggesting the presence of cartilage padding that would help with the flicking. The tail also seemed to lash horizontally rather than vertically, which made sense given the design of the vertebrae and the supporting structures around them.

If the animal whipped the base of its tail sideways at a speed of something like 6 1/2 feet (2 meters) per second, that would magnify to propel the tip over the speed of sound. And it would require less energy than walking! Given that the animal carried 70 percent of its weight in its hindquarters, simply stamping its feet could send the tail into a lash. So what would a dinosaur tail breaking the sound barrier sound like? Myhrvold and Currie calculated a noise that might have reached two hundred decibels—enough to startle even a *T. rex*!

There were critics of Myhrvold and Currie's theory, of course, and in response Myhrvold built a replica of the hindquarters and tail of the *Apatosaurus*. It is quarter-sized, weighing 44 pounds (20 kilograms) with a length of merely 11 feet (3.5 meters), but it is exactly scaled to the tail of the dinosaur, complete with eighty-two vertebrae constructed of steel, neoprene, aluminum and Teflon. When you crank the tail, the tip breaks the sound barrier and creates that satisfying crack, suggesting that it may well have been possible that this dino broke the sound barrier 150,000,000 years ago. So it's not a tall tale about a long tail, but rather as Currie and Myhrvold called it: supersonic sauropods.

How many times can you fold a piece of paper?

IT IS SAID THAT YOU CAN FOLD A PIECE OF PAPER—any size paper—only about eight times. After that, the resistance is too strong, and there's often no room left for folding. So is that true? What if you use a larger sheet or a thinner sheet? What if a more dexterous person does the folding? As it turns out, there is a limit on folds, no matter how you try to get around it, but the magic number of eight might not be so magic after all.

Many people have tried to get around the fold problem. The mathematics world owes a debt to Britney Gallivan, high school student extraordinaire. In 2002, she took on the fold problem for extra credit in her math class. She first worked out an equation that suggested more than eight folds were at least theoretically possible, then proceeded to demonstrate that was true by folding not a piece of paper but a four-by-four-inch (ten-by-ten centimeter) sheet of

gold foil. (The foil was 0.3 micrometers thick, or more than three hundred times thinner than the typical piece of paper.)

Britney's task wasn't easy: a piece of gold that thin is extremely prone to tearing, and she had to work painstakingly with artist's brushes and tweezers to fold the sheet without having it disintegrate in the process. But she did it, folding the foil an inspiring twelve times! She handed in her results—only to be told by her teacher that the extra credit was specifically for folding a piece of paper, not a piece of gold. Since she had solved a different problem, she wasn't going to receive the additional credit. (Aren't teachers supposed to support kids who go a little further?) Most kids her age (or adults, for that matter) would have abandoned the problem, but not Britney.

She came back armed with paper—an unrolled sheet of toilet paper, to be exact, 4,000 feet (1.2 kilometers) long—and she folded it in half twelve times. Her teacher again complained about her results, this time saying the folds on the final product looked too rounded and so her solution wasn't definitive.

But here's what her teacher missed: Britney's analysis essentially showed that as folds inevitably become rounder and rounder, a limit is indeed reached. At some point, there just isn't enough unfolded paper to do the wrapping. Try it yourself and you'll see.

 TRY THIS: To try the folding experiment yourself, start with a typical sheet of 8 1/2- by 11-inch printer paper. Fold it in half along its length, then in half again along its width. Fold as tightly and neatly as you can, creating a knife-edge crease. After a mere four folds, you will notice the resistance has turned from trivial to substantial. You'll also notice that as the number of folds increases, the geometric precision is lost, and folds gradually turn into bends as the thickness makes sharp creases more and more difficult. So how many clean folds did you make? Five or six, perhaps? After five or six folds, there is no way your bare hands can keep folding against the resistance. Now try the same technique with a sheet of 11- by 17-inch paper. Fold it the same way described above. You'd think because of the increased size that you'd be able to get more folds out of this sheet, but that's not so. How many does it take before the resistance is too strong? Five, maybe? Six? So much for the theory that you can fold a piece of paper about eight times. In truth, it's much less.

Britney's achievement wasn't underappreciated only by her teacher: a paper by Professor Graham Rees in the journal *Philosophical Magazine Letters* three years later overlooked her work completely, claiming that the usual number of folds was six, and in extreme cases, eight. When Rees was informed that Britney's experiment had proved him wrong, he admitted that he wasn't aware of her and had intended his paper as a spoof anyway. (Lesson: if you're an academic, try looking online to see whether anyone has scooped you, and not just in the academic journals.)

Is Britney's discovery—that the thinnest sheet can be folded twelve times and after that the paper is too rounded and small to fold—the definitive answer to this puzzle? Probably, for anything practical. But that doesn't mean we shouldn't dream.

What if we use even thinner sheets than Britney's gold foil? Paper is mostly cellulose, but there is something new called nanocellulose, a super-micro version of the common cellulose molecule that has extraordinary properties. It has the strength of Kevlar but is incredibly light. More important, being a nanomaterial, nanocellulose can be manipulated on the atomic level, and there really isn't anything that could be thinner than that.

I posed the question of how many times a nanomaterial could be folded to scientists at the National Institute for Nanotechnology in Edmonton, Alberta. Grad student Roshan Achal was intrigued enough to tackle the problem. He chose to consider graphene, a honeycomb-like sheet of carbon atoms that is just one atom thick. Graphene is not nanocellulose, but scientists have a better understanding of it.

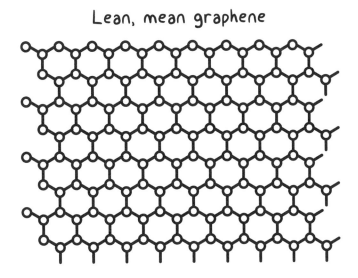

Lean, mean graphene

Also, graphene can be made into sheets about 0.3 nanometers thick, a thousand times thinner than Britney's gold foil and a million times thinner than the standard piece of paper. (It should be said that making large sheets like this is still incredibly challenging, and operating on the nano level means that electrical and chemical effects, essentially absent when using paper, might affect the experiment.)

Roshan plugged the thickness of the graphene sheet into the equations to discover—at least theoretically—that graphene can be folded nineteen times! Further, Britney's twelve-fold record was set using a continuous roll of toilet paper nearly three-quarters of a mile long (more than a kilometer). If you could make a graphene sheet that big, you might be able to fold it a lot more than nineteen times—in fact, more like twenty-seven times. Not that you would actually ever go to that trouble. Or would you?

This plan is unraveling.

There's a difference between this theoretical limit of graphene and Britney's toilet-paper-folding experiment: Roshan's method was to fold, turn the graphene ninety degrees, fold, turn ninety degrees and so on, whereas Britney had folded her paper in the same direction every time. Plugging graphene into the equations for her method gives slightly different results: a sheet of graphene the same size as a piece of printer paper could be folded twelve times. If that graphene were then made into a sheet the length of Britney's toilet paper, it could be folded twenty-one times—close to the limit that Roshan found.

One final point: Roshan observes that if you started with a sheet one atom thick and folded it nineteen times, you'd actually be able to see it after the folds were made. That's because the height of the accumulated folds is growing exponentially. It is startling enough that you could see a stack of graphene atoms with the naked eye after just a few folds, but here are some equivalent examples using paper that make the same point much more dramatically:

- Folding an average piece of paper ten times would create a stack the thickness of a hand.
- Twenty-three folds would stand about two-thirds of a mile (a kilometer) high. Another seven would take you up 62 miles (100 kilometers)—into outer space.
- Fifty-one folds would take you plunging into the sun.
- And 103 folds—clearly impossible, mind you—takes you to the edge of the observable universe. Or beyond.

Origami can be dangerous.

How are champagne bubbles different from beer bubbles?

ALL BUBBLES HAVE A LIFE THAT IS—by human standards, anyway—extremely short. Their deaths are even shorter. Bubbles in both beer and champagne contain pressurized carbon dioxide (CO_2). During the fermentation process of both drinks, yeast breaks down sugar, generating, among other things, alcohol and carbon dioxide. As the carbon dioxide is created, it dissolves into the beer or champagne, and when the bottle is sealed with a cap or cork, the gas is trapped inside. Some carbon dioxide collects in the airspace just under the cap or cork, but the rest remains dissolved. They're in balance as long as the cap or cork is in place, and so bubbles will not form. But when the seal is broken, the pressure in the airspace drops and the gases, including the carbon dioxide, rush out of the bottle. Now the balance is disturbed, and the rest of the CO_2 inside the drink must escape. The most efficient way for the CO_2 in the liquid to do that is in the form of a bubble.

For bubbles to form and escape, chemistry demands that there be "nucleation sites"—places where the bubbles can settle and grow—on the inside of the glass. It used to be thought that these were mostly defects in the glass surface, but further research has shown that they are more likely to be tiny stray fibers of cellulose that have settled in the glass from the air or a cloth.

Even when a nucleation site is available, though, forming a bubble is a tricky business. After a bottle is opened, CO2 molecules rush to each nucleation site. When a bubble is first developing, it's less than 1/100,000 of an inch across, but as more carbon dioxide molecules collect together, they eventually reach a critical size, become buoyant enough and start to rise in a bubble. As each bubble floats upward, it becomes its own nucleation site, and so it collects more and more CO2 as it rises. As it ascends rapidly to the surface of the liquid, it swells to a couple of thousand times its original size.

When a bubble reaches the surface, most of it remains submerged, but a small part of it protrudes into the air. The liquid of that exposed cap starts to drain away, to the point where almost anything—a slight tilt of the glass or bottle, even a wisp of air passing by—will burst it. When a bubble breaks, the impact causes a high-speed jet of beer or champagne to explode upward from the surface. Multiply that by hundreds of bubbles and you can't miss the popping and fizzing of the tiny alcoholic fireworks display. When you settle in for a sip, some of those flying bubble fragments bring the flavors of your favorite drink to your nose—the more broken bubbles, the more intense the aroma and flavor.

 TRY THIS: Open a bottle of beer or champagne and then watch the strings of bubbles as they float to the surface. You'll notice that the bubbles in the string are farther apart the closer they get to the surface of the liquid. That's because the more CO2 that the bubble collects, the more buoyant it becomes and the faster it moves.

There are many differences between the bubbles in champagne and those in beer. For one thing, a bottle of champagne contains three to four times as much CO2 as a bottle of beer. Also, a bubble release site in champagne can generate up to thirty bubbles a second, roughly three times the number in beer. That explains the fun of popping the cork off a bottle of champagne in a celebration. Because there is a greater concentration of bubbles in champagne—they grow faster, and the bottle is bigger than a typical beer—the "eruption velocity" of champagne is one hundred times that of a bottle of beer.

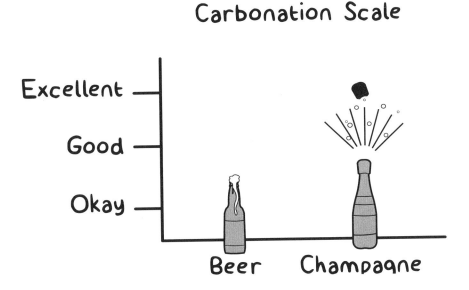

Both beer and champagne are complex mixes of elements, and some of those chemicals can attach to the surface of a bubble. This happens more in beer than it does in champagne. In beer, some substances—for instance high-molecular-weight proteins and isohumulones from the hops—stabilize the bubbles, slowing their upward mobility and making them less likely to burst.

As the bubbles in beer hit the surface, instead of breaking apart, they bind together and form a frothy head. Champagne is very different: there are fewer of these stabilizing substances, and the bubbles rise much faster, ensuring that champagne bubbles burst immediately upon reaching the surface.

When it comes to pouring the perfect head of foam on a beer, Guinness reigns supreme (and no, they didn't pay me to say that). The head on a pint of Guinness is smoother and far longer lasting than it is on other beers. The secret? Guinness contains nitrogen gas in addition to CO_2. Bubbles of nitrogen are different from regular old carbon dioxide: they're smaller and better at supporting the kind of dense foam that characterizes a glass of Guinness.

In the pub, a Guinness pour—or "pull"—is forced under pressure through something called a creamer plate, which is filled with tiny holes that minimize the size of the bubbles. But when Guinness began to sell their beer in cans, re-creating those same conditions was a challenge. Nitrogen doesn't dissolve willingly in beer, so most of the gas—which was introduced into the liquid during the canning stage—hung around at the top and escaped when the can was popped. No nitrogen in the liquid meant no distinctive Guinness head.

Guinness solved that problem by introducing the "widget": a tiny ball filled with nitrogen and carbon dioxide that's added to each can. As long as the can is sealed, the widget doesn't do anything. But when the can is opened, the pressure change causes the widget to release its combination of gases. The stream of nitrogen bubbles agitates the beer, creating a long-lasting creamy head.

There's one final oddity about Guinness. If you watch a glass of the stuff carefully, you are likely to see streams of bubbles descending, not ascending the way bubbles in champagne do. That is weird, because bubbles are buoyant, and so they should always rise in beer. The reason they don't in Guinness is that the bubbles are so small. The smaller the bubbles, the less buoyant they are, and so the more they're affected by other forces. In the case of Guinness, that "other force" is an upward flow of liquid in the middle of the glass. As the carbon dioxide bubbles rise up from the bottom of the glass, they drag liquid along with them, and where there's rising liquid there has to be falling liquid—or else the beer would levitate out of the glass!

In a traditional pint glass, the carbon dioxide escapes up the middle of the glass, and so the nitrogen bubbles—because they're too weak to resist other forces—are pushed to the outer edges of the glass and descend along the wall. In one cool experiment, though, researchers showed that an "anti-pint" glass—one that's fatter at the bottom than the top—can reverse these currents. If Guinness is poured into an anti-pint glass, the bubbles flow downward in the middle of the glass and rise along its walls. The experimenters came to a simple conclusion: because the anti-pint glass is wider at its base, carbon dioxide bubbles that form along the bottom are crowded together by the walls as they rise. The only place the nitrogen can sink is right down the middle, between the escaping streams of CO_2 bubbles.

If you want to try these experiments yourself, Guinness is the best beer to use because the sinking bubbles are highlighted against the beer's dark color. It makes for easy viewing . . . but remember, if you tilt the glass to drink from it, you're going to disturb the bubble display.

Why does the moon look larger at the horizon?

If you look at the moon when it is just rising or setting, it looks almost twice as big as it does when it's overhead. Although people have seen this happen with every full moon for millennia, it has been extremely difficult to come up with an explanation for this optical illusion.

Early stargazers suggested that the moon actually changed size. They believed it inflated during the day, like a balloon, out of our sight, then rapidly shrank as it rose in the night sky, only to reinflate just before dawn. But that theory is (justifiably) long gone.

It is true that because the moon's orbit is more of an oval than a circle, at any given point in its orbital cycle its distance to the earth can vary by more than 24,000 miles (40,000 kilometers). The change in the distance between the moon and earth, however, takes more than a single night to happen. So that doesn't explain why the moon looks so much bigger at dawn or dusk.

On to the next: people reasoned the moon might appear larger at the horizon because it was closer to the earth then. The opposite is actually true, although by an unimportant amount: when you look at the overhead moon, you're standing on the closest point on the earth to it, but the horizon moon is literally half an earth (roughly 3,700 miles, or 6,000 kilometers) farther away.

Another possibility was then considered: that the atmosphere somehow bent the light that reflected off the moon. But you can quickly disprove that idea with photographs. Sequential pictures of the moon taken over the course of a night show that it's exactly the same size at all times—that is, the size of the moon in a photograph will always be the same no matter where in the sky the moon is when you take the picture, unlike the apparent dramatic changes we see when we're looking at it with our eyes.

 TRY THIS: You can do a simple test to check the size of the moon and see for yourself whether the change in size is an illusion. Take an aspirin outdoors when there's a full moon, and hold it up at arm's length to cover the face of the moon. First, you'll be astonished to see that a tiny aspirin can cover that giant horizon moon, and you'll be even more astonished to see that later that night, when the moon is overhead, it will still just cover it. Next, observe the horizon moon through a toilet paper roll. The moon will suddenly appear smaller than it did when you were looking at it without the cardboard tube. What the photograph and toilet paper roll experiments show is that the differing sizes of the moon appear to have something to do with how we view it. More accurately, the difference has to do with how we think we see the moon. It's all an illusion.

Although the moon's varying sizes are an illusion, distance still plays a key role in how it appears to us. Your distance to the moon is more or less the same no matter where the moon is in the sky. You do get closer to the moon as the earth turns, but as explained earlier, that distance is trivial (relatively speaking). What's more important than the distance itself is how far that distance seems to our brains. One theory is that the horizon moon seems much farther away than the overhead moon because you can compare the horizon moon to everything on the ground between you and it—such as buildings, mountains or trees.

The problem with this theory is that we don't really need any intervening objects to experience the moon illusion. Pilots at high altitudes have experienced the bigger horizon moon, even with no other large or moving objects to compare their view against. The moon illusion also works over deserts or water—places where there are few, if any, geographical features.

 TRY THIS: On a clear evening, turn your back on the horizon moon, then bend down and look at it from between your legs. You will see that the moon appears to be smaller from this awkward vantage point. You can even try an experiment within this experiment: How far upside down do you have to be before the moon looks small?

So what is it that makes the moon look big near the horizon? We know that the image the moon leaves on your retina is the same all night, from moonrise to moonset; its size never changes. One of the most popular explanations is that your brain doesn't imagine the sky to be the inside of a sphere, but rather, more like a shallow inverted bowl —distant at the horizon but much closer overhead. So your brain has two conflicting inputs.

On one hand, the image of the moon on your retina is the same at every point, but your brain has to reconcile that with its assumption that the horizon moon is actually farther away. The only way two objects at different distances away can leave equal-sized images in your eye is if one is bigger. How can it make the same size image if it's farther away? It must be bigger! And so it appears that way.

History Mystery

Did Galileo drop balls from the Tower of Pisa?

A POPULAR LEGEND RELATES HOW, sometime in the late 1580s, Galileo climbed the Leaning Tower of Pisa carrying a musket ball and a cannonball. He was there to wage war—not on any army, mind you, but on two thousand years of scientific belief.

Heads up!

Up to that point, much of physics had been based on the teachings of Aristotle. One of the ancient Greek's beliefs was that objects fell in proportion to their weight. Aristotle claimed that if a 100-pound lump of clay and a 1-pound lump of clay were dropped at the same time, the bigger lump would fall 100 feet for every 1 foot the smaller one fell. Seems ludicrous now, but that was the prevailing belief then.

Galileo didn't set out that day to prove that anything of any size, weight or density would fall at the same speed—he didn't believe that. What he did believe was that if he dropped a musket ball and a cannonball together, they would both hit the ground at virtually the same time. As a group of students, teachers and other scientists (then called philosophers) watched from the ground with bated breath, Galileo supposedly dropped the two balls over the ramparts of the tower. It was a long way down to the ground, but the story goes that both balls hit the grass simultaneously, a brilliant confirmation of Galileo's claims and an utter rejection of Aristotle's theory, which had held sway for close to two thousand years.

If it happened today, tweets, Instagrams and animated GIFs would dominate social media. There'd be video footage. And there would be no doubt about how the events took place. But this was more than four hundred years ago. There was no photography or even an artist to sketch the event. Actually, nobody at the time even wrote about what happened. It wasn't until sixty years later—twelve years after Galileo died—that the experiment was recounted for the first time. The entire legend was captured in a few scant lines written by Galileo's close friend and biographer Vincenzo Viviani. Viviani suggested that Galileo had performed the experiment more than once, and several writers since have embellished the story to the point that they claimed the onlookers watched with horror as their cherished faith in Aristotle was demolished before their eyes. (The YouTube videos would certainly have included interviews with them.)

Considering the intensity of the arguments waged over whether the experiment at Pisa even happened, it's fair to ask: Who cares? After all, it doesn't matter how exactly Galileo arrived at his findings, just that he did so. But there's good evidence that at least two others before Galileo might have beaten him to the punch. In the seventh century, a little-known individual named John Philoponus said that if you dropped two objects of very different weights, they would fall, not in the ratio of those weights (the Aristotle line of thinking), but at the same time. According to Philoponus, even if one of the objects was double the weight of the other, the difference in the rate of their fall would be imperceptible.

Later, a mere four years before Galileo was said to have climbed the Tower of Pisa, a mathematician and engineer named Simon Stevin, in Delft, the Netherlands, described how he had dropped two iron balls—one ten times heavier than the other—onto a plank thirty feet (nine meters) below. Stevin reported that the two balls hit the plank so simultaneously that it sounded like a single impact.

Galileo himself never breathed a word about demonstrations at the Tower of Pisa (except, allegedly, to Viviani). He did, however, claim to have "made the test" with a cannonball and a musket ball, after which he began to predict what would happen if he dropped other objects. For instance, he predicted that if a lead ball and an ebony ball were dropped from a height of about 330 feet (100 meters), they would be no more than about a finger length apart when they hit the ground. (Note: he didn't say that he had actually tried it.)

Despite his celebrated (purported) demonstration and the fact that Galileo dragged science out of ancient Greece and into the seventeenth century, one of the ironies of his work is that his predictions about the falling objects were incorrect. Galileo neglected to consider air resistance. Objects falling through air experience drag caused by friction. The greater the velocity, the greater the drag. If something is falling from high enough, such as a parachutist, that person will accelerate until air resistance increases to equal the force of gravity, and at that point the parachutist will have reached terminal velocity.

The degree to which air resistance affects a falling object depends on both the size and density of that object. From the 1960s to the 2000s various attempts were made to re-create the Pisa experiment.

In one, an iron ball and a rubber ball of the same size (but obviously very different densities) were dropped 125 feet (38 meters)—the equivalent of a thirteen-story building—and when the iron ball hit the ground, the rubber ball was about 23 feet (7 meters) behind it. Scientists have also calculated what really would have happened with Galileo's conjectured lead versus ebony: if both balls were about 4 inches (10 centimeters) in diameter and were dropped 330 feet (100 meters), upon impact they would be 16 feet (5 meters) apart, not a finger length, as Galileo had predicted. Ebony and lead are different densities, but even if you take that out of the equation, the great scientist was still off. Galileo had theorized that if two iron balls one weighing a pound and the other 100 pounds, were dropped a little more than 200 feet (60 meters), there would be only two inches (five centimeters) between they when they hit the ground. The reality is that they would be more than 4 feet (1.2 meters) apart.

In the absence of air friction, all kinds of combinations of sizes, densities and shapes of objects would fall at identical speeds and land at the same time. That's been proven by performing tests in a vacuum. While it's possible to create a vacuum on Earth (and the Galileo experiment has been duplicated in one), it's far more spectacular to test the theory on the moon. When the Apollo 15 astronauts were there, astronaut David Scott performed the experiment with a feather and a hammer. He dropped both at the same time, and the grainy video of the test shows both objects landing on the moon's surface at exactly the same time. Although the astronauts were confident of the results they would get, there were apparently risks that the experiment wouldn't work. Before the successful trial, Scott had tried it once, and static electricity caused the feather to stick to his hand. Scott's bad luck revealed another important issue involved when trying to drop two objects of very different weights simultaneously while on the moon or anywhere: you tend to let go of the lighter one first, especially if you've been holding them at arm's length.

TRY THIS: Cut out a paper disk the same size as a quarter. Put the paper disk on the tip of one forefinger and an actual quarter on the tip of the other. Now drop both of them simultaneously. The coin will reach the floor before the paper disk. From this experiment, it is possible to conclude—mistakenly—that heavier objects fall faster.

Now place the paper disk on top of the coin and drop them together. Both will reach the floor at the same time. Why? By putting the paper on top of the coin, you've shielded it from the air resistance it encountered in the first experiment. The second experiment shows that it's not the mass of the object that causes it to fall faster or slower but the resistance or air friction.

It's possible that none of these facts have convinced you and that you still think that heavier objects fall faster than lighter ones. It's an intuitive conclusion, but intuitive physics leads people astray all the time. After all, about a quarter of North Americans believe that if a person runs off a cliff, they'll move straight out until they slow down and stop, at which point they'll fall straight down à la Wile E. Coyote. If you're of that mindset, then consider this thought experiment. Pretend we have a ten-pound ball and a one-pound ball. Let us assume, à la Aristotle, that the ten-pound ball falls faster than the one-pound ball because it's heavier. Now, let's attach the two balls. What happens next? A traditional Aristotelian might think that the combined object should fall slower, since the slower movement of the one-pound ball would hold back the ten-pound ball's descent. But by the same group's reasoning, the two balls should fall faster when they're attached because they weigh more, and heavier objects move faster. It's impossible for both outcomes to happen, so the only possibility is that they were falling at the same rate in the first place. Chalk up another point for Galileo.

It's very unlikely that further documentation will surface to shed definitive light on the incident at Pisa. So why, as I asked before, is something so apocryphal so celebrated? In a single, dramatic story, one man completely reversed millennia-old thinking about how things work. It's not the science; it's the showmanship!

That's why.

Acknowledgments

There I was, wondering if I would ever write another book, when my old friend Kevin Hanson sprung one on me. Of course I accepted, and this is the result.

Nita Pronovost, with Brendan May, shepherded the manuscript from beginning to end, although there might have been times when it felt like the shepherds were flogging the sheep (me). At any rate, it is always true that another set of eyes on a manuscript makes it better, and whenever Nita questioned some of my brilliant prose, she was right. She made me think: "How could I have written that?"

I'm looking forward to working with the Simon & Schuster publicity team, Catherine Whiteside et al., because I know that they will make the post-operative malaise a whole lot better.

My agent, Jackie Kaiser, and Jake Babad at Westwood Creative Artists have made the whole process run smoothly. I have to say, writing with an agent is a whole lot better than writing without one.

I had help on this book from two fantastic researchers both of whom I know from the Banff Science Communications Program at the Banff Centre. Niki Wilson, an accomplished science writer on her own, and Joanne O'Meara, a physics prof at the University of Guelph and a great communicator too, took on the job of scouring the scientific literature on biology (Niki) and physics (Joanne). They were great and, given that there was a tight timeline on this book, essential. I'm glad I put my faith in them.

Finally, I'm lucky to have a set of inquisitive friends who will tolerate my talking (babbling) about this or that scientific. They include Wendy Tilby, Amanda Forbis, Dave Nicholls, Denise Jakal, Trevor Day, Garth Kennedy, Ben Jackson, Steve Dodd, Josip Vulic and Penny Park.

Mary Anne, of course, serves as yet another set of eyes and—way more important—ongoing support.

Thanks to all.

Photo: Richard Siemens

Jay Ingram was the host of Discovery Channel Canada's *Daily Planet* from the first episode until June 2011. Before joining Discovery, Ingram hosted CBC Radio's national science show, *Quirks & Quarks*. He has received the Sandford Fleming Medal from the Royal Canadian Institute, the Royal Society of Canada's McNeil Medal for the Public Awareness of Science and the Michael Smith Award for Science Promotion from the Natural Sciences and Engineering Research Council of Canada. He is a distinguished alumnus of the University of Alberta, has received five honorary doctorates and is a Member of the Order of Canada. He has written thirteen books, including many bestsellers. Visit Jay at **JayIngram.ca**.

🐦 @jayingram